T0271365

Internet of Things and Big Data Analytics-Based Manufacturing

By enabling the conversion of traditional manufacturing systems into contemporary digitalized ones, Internet of Things (IoT) adoption in manufacturing creates huge economic prospects through reshaping industries. Modern businesses can more readily implement new data-driven strategies and deal with the pressure of international competition thanks to Industrial IoT. But as the use of IoT grows, the amount of created data rises, turning industrial data into Industrial Big Data.

Internet of Things and Big Data Analytics-Based Manufacturing shows how Industrial Big Data can be produced as a result of IoT usage in manufacturing, considering sensing systems and mobile devices. Different IoT applications that have been developed are demonstrated, and it is shown how genuine industrial data can be produced, leading to Industrial Big Data. This book is organized into four sections discussing IoT and technology, the future of Big Data, algorithms, and case studies demonstrating the use of IoT and Big Data in a variety of industries, including automation, industrial manufacturing, and healthcare.

This reference title brings all related technologies into a single source so that researchers, undergraduate and postgraduate students, academicians, and those in the industry can easily understand the topic and further their knowledge.

Intelligent Manufacturing and Industrial Engineering

Series Editor: Ahmed A. Elngar Beni-Suef Uni.
Mohamed Elhoseny, Mansoura University, Egypt

Machine Learning Adoption in Blockchain-Based Intelligent Manufacturing
Edited by Om Prakash Jena, Sabyasachi Pramanik, and Ahmed A. Elngar

Integration of AI-Based Manufacturing and Industrial Engineering Systems with the Internet of Things
Edited by Pankaj Bhambri, Sita Rani, Valentina E. Balas, and Ahmed A. Elngar

AI-Driven Digital Twin and Industry 4.0: A Conceptual Framework with Applications
Edited by Sita Rani, Pankaj Bhambri, Sachin Kumar, Piyush Kumar Pareek, and Ahmed A. Elngar

Technology Innovation Pillars for Industry 4.0: Challenges, Improvements, and Case Studies
Edited by Ahmed A. Elngar, N. Thillaiarasu, T. Saravanan, and Valentina Emilia Balas

Internet of Things and Big Data Analytics-Based Manufacturing
Edited by Arun Kumar Rana, Sudeshna Chakraborty, Pallavi Goel, Sumit Kumar Rana, and Ahmed A. Elngar

For more information about this series, please visit: https://www.routledge.com/ Mathematical-Engineering-Manufacturing-and-Management-Sciences/book-series/ CRCIMIE

Internet of Things and Big Data Analytics-Based Manufacturing

Edited by
Arun Kumar Rana,
Sudeshna Chakraborty, Pallavi Goel,
Sumit Kumar Rana, and Ahmed A. Elngar

CRC Press
Taylor & Francis Group
Boca Raton London New York

CRC Press is an imprint of the
Taylor & Francis Group, an **informa** business

Designed cover image: Shutterstock – greenbutterfly

MATLAB® and Simulink® are trademarks of The MathWorks, Inc. and are used with permission. The MathWorks does not warrant the accuracy of the text or exercises in this book. This book's use or discussion of MATLAB® or Simulink® software or related products does not constitute endorsement or sponsorship by The MathWorks of a particular pedagogical approach or particular use of the MATLAB® and Simulink® software.

First edition published 2025
by CRC Press
2385 NW Executive Center Drive, Suite 320, Boca Raton FL 33431

and by CRC Press
4 Park Square, Milton Park, Abingdon, Oxon, OX14 4RN

CRC Press is an imprint of Taylor & Francis Group, LLC

ISBN: 978-1-032-66671-6 (hbk)
ISBN: 978-1-032-67346-2 (pbk)
ISBN: 978-1-032-67347-9 (ebk)

DOI: 10.1201/9781032673479

Typeset in Times
by codeMantra

Contents

SECTION 1   IoT and Technology

SECTION 2   The Future of Big Data

SECTION 3   Algorithms

SECTION 4 Case Studies/Uses of IoT and Big Data in Different Domains

Preface

Welcome to *Internet of Things and Big Data Analytics-Based Manufacturing*, an edited volume that delves into the transformative fusion of two groundbreaking technologies – the Internet of Things (IoT) and Big Data analytics – within the realm of manufacturing. This book explores how the convergence of these technologies is reshaping traditional manufacturing processes, driving efficiency, innovation, and competitiveness in today's industrial landscape.

The Internet of Things, with its interconnected network of physical devices embedded with sensors, actuators, and software, has revolutionized the way machines communicate, collect data, and interact with their environment. Concurrently, the exponential growth of data generated by these IoT devices has provided a wealth of information that holds immense potential for businesses. Big Data analytics, equipped with advanced algorithms and tools, has the power to extract valuable insights from this data deluge, enabling manufacturers to optimize operations, enhance product quality, and create new revenue streams.

In this book, researchers from academia and industry come together to explore the multifaceted impact of IoT and Big Data analytics on manufacturing. Through a collection of insightful chapters, readers will gain a comprehensive understanding of the key concepts, methodologies, applications, and challenges associated with leveraging these technologies in various manufacturing domains. By providing insights into the latest advancements and best practices, this book serves as a valuable resource for researchers, practitioners, engineers, and students interested in harnessing the potential of IoT and Big Data analytics for revolutionizing manufacturing processes.

We extend our sincere gratitude to all the contributors who have shared their expertise and insights to make this edited volume possible. We hope that this book will inspire further exploration, innovation, and collaboration in the dynamic intersection of IoT, Big Data analytics, and manufacturing.

About the Editors

Arun Kumar Rana has completed his B.Tech. degree from Kurukshetra University, and M.Tech. and Ph.D. degrees from Maharishi Markandeshwar (Deemed to be University), Mullana, India. His area of interest includes Image Processing, Wireless Sensor Network, Internet of Things, AI, and Machine Learning and Embedded systems. Prof. Rana is currently working as an Assistant Professor at Galgotias College of Engineering and Technology, Greater Noida, India, with more than 16 years of experience. Prof. Rana is a collaborative researcher. He has published 25 SCI papers (6 submitted), 30 SCOPUS papers, 25 chapters, and 30 papers in national and international conferences. He has also published 10 books with a national and international publisher like Taylor and Francis (Scopus Index), USA.

Sudeshna Chakraborty has 18.5+ years of experience in almost all Academic and Industry-Academia Interface Processes including teaching UG, PG and Ph.D. scholars with the remarkable role of Exam Department Supervisor. She has published 18 national and international Patent. She has Publications in 50+ national international journals/Conferences. She has worked on a Funded Project of Rs 15 Lacson E-learning and Brain–Computer Interface SHe is the recipient of Outstanding Academician Award 2020, Research Excellence Award, Academician's Award by Institute of Scholars 2021, Nikhil Bhartiya Parishad Shiksha Parishad 2022, and Best Speaker Award on Roles and Trends of Engineers on Engineering Day by Institute of Engineers of India.

Pallavi Goel is currently a Professor at the CSE Department Faculty of Engineering and Technology, also leading various roles at the departmental level (like industry collaboration and government-funded projects). He is in charge of IPR for Galgotias College of Engineering and Technology, Greater Noida, India. Dr Pallavi Goel was an Associate Professor at the CSE Department Faculty of Engineering and Technology and international research collaboration coordinator for Manav Rachna International Institute of Research and Studies, Surajkund, Delhi. She worked for nine years for the School of Computing Science and Engineering at Galgotias University.

Sumit Kumar Rana has completed his B.Tech. degree from Kurukshetra University, and M.Tech. and Ph.D. degrees from Maharishi Markandeshwar (Deemed to be University), Mullana, India. His areas of interest include blockchain technology, cryptography, cryptocurrency, and artificial intelligence. Prof. Sumit Kumar Rana is currently working as an Assistant Professor at Panipat Institute of Engineering and Technology, Panipat, India, with more than 12 years of experience. Prof. Rana is a collaborative researcher. He has published multiple SCI/SCOPUS papers, book chapters and papers at national and international IEEE conferences.

Ahmed A. Elngar is an Associate Professor and Head of the Computer Science Department at the Faculty of Computers and Artificial Intelligence, Beni-Suef University, Egypt. Dr. Elngar is also Associate Professor of Computer Science at the College of Computer Information Technology, American University in the Emirates, United Arab Emirates. Also, Dr. AE is an Adjunct Professor at the School of Technology, Woxsen University, India. Dr. AE is the Founder and Head of the Scientific Innovation Research Group (SIRG). Dr. AE is a Director of the Technological and Informatics Studies Center (TISC), Faculty of Computers and Artificial Intelligence, Beni-Suef University. Dr. AE has more than 150 scientific research papers published in prestigious international journals and over 35 books covering such diverse topics as data mining, intelligent systems, social networks, and smart environments. Dr. AE is a collaborative researcher. He is a member of the Egyptian Mathematical Society (EMS) and the International Rough Set Society (IRSS). His other research areas include the Internet of Things (IoT), Network Security, Intrusion Detection, Machine Learning, Data Mining, Artificial Intelligence, Big Data, Authentication, Cryptology, Healthcare Systems and Automation Systems. He is an editor and reviewer of many international journals around the world. Dr. AE won several awards including the "Young Researcher in Computer Science Engineering" from the Global Outreach Education Summit and Awards 2019, on 31 January 2019 (Thursday) in Delhi, India. Also, he was awarded the Best Young Researcher Award (Male) (Below 40 Years) and Global Education and Corporate Leadership Awards (GECL-2018)

Contributors

Ashima Arya
KIET Group of Institutions
Ghaziabad, India

Ashok
Government P.G. College
Ambala, India

Anupam Kumar Bairagi
Khulna University
Gollamari, Bangladesh

Preeti Bansal
Chandigarh Group of Colleges
Mohali, India

Ravi Kumar Barwal
Government College for Women
Shahzadpur, India

Pranto Bosu
Guru Nanak Dev University
Amritsar, India

Nidhi Chahal
Chandigarh Group of Colleges
Mohali, India

Sachin Chaudhary
IIMT University
Meerut, India

Anita Dahiya
Panipat Institute of Engineering and
    Technology
Panipat, India

Biva Das
Khulna University
Gollamari, Bangladesh

Padam Dev
O'Chicken India
Chandigarh, India

Mamta Devi
CT Institute of Technology and
    Research
Jalandhar, India

Ritu Dewan
Lingaya's Vidyapeeth
Faridabad, India

Mohit Dua
National Institute of Technology
Kurukshetra, India

Shelza Dua
Punjab Engineering College
Chandigarh, India

Alok Dutta
Adamas University
Kolkata, India

Debosree Ghosh
Shree Ramkrishna Institute of Science
    and Technology
Sonarpur, India

Vineeta Gulati
Maharishi Markandeshwar (Deemed to
    be University)
Ambala, India

Prince Gupta
KIET Group of Institutions
Ghaziabad, India

Sapna Juneja
KIET Group of Institutions
Ghaziabad, India

Amandeep Kaur
Chitkara University
Rajpura, India

Gaganpreet Kaur
Chitkara University
Rajpura, India

Mandeep Kaur
Maharishi Markandeshwar
    (Deemed to be University)
Ambala, India

Simarpreet Kaur
Chandigarh Group of Colleges
Mohali, India

Amit Kumar Kesarwani
Galgotias College of Engineering and
    Technology
Greater Noida, India

Bhupendra Kumar
IIMT University
Meerut, India

Naveen Kumar
Mahatma Jyotiba Phule Rohilkhand
    University
Bareilly, India

T. S. Pradeep Kumar
Vellore Institute of Technology
Chennai, India

Lavisha
Maharishi Markandeshwar
    (Deemed to be University)
Ambala, India

Shresth Modi
Chandigarh Group of Colleges
Mohali, India

Anita Mohanty
Silicon Institute of Technology
Bhubaneswar, India

Subrat Kumar Mohanty
Einstein Academy of Technology and
    Management
Bhubaneswar, India

Ambarish G. Mohapatra
Silicon Institute of Technology
Bhubaneswar, India

Anindya Nag
Khulna University
Gollamari, Bangladesh

Tapsi Nagpal
Lingaya's Vidyapeeth
Faridabad, India

Sasmita Nayak
Government College of Engineering
Kalahandi, India

Rajneesh Panwar
IIMT University
Meerut, India

Pratibha
Chitkara University
Rajpura, India

Abhishek R. A.
Vellore Institute of Technology
Chennai, India

Renuka Devi Rajagopal
Vellore Institute of Technology
Chennai, India

Arun Kumar Rana
Galgotias College of Engineering and
    Technology
Greater Noida, India

Sumit Kumar Rana
Panipat Institute of Engineering &
    Technology
Panipat, India

Kumar Rangasamy
Vellore Institute of Technology
Chennai, India

Manju Rani
Gurugram University
Gurugram, India

Giri Sainath Reddy
National Institute of Technology
Kurukshetra, India

Neeraj Rohilla
Government College
Barwala, India

Himanshi Sachan
Chandigarh Group of Colleges
Mohali, India

Jaya Sharma
KIET Group of Institutions
Ghaziabad, India

Kewal Krishan Sharma
IIMT University
Meerut, India

Vikas Sharma
IIMT University
Meerut, India

Riya Sill
Adamas University
Kolkata, India

Inderpal Singh
CT Institute of Technology and
    Research
Jalandhar, India

Shikha Singh
Galgotias College of Engineering and
    Technology
Greater Noida, India

Swasti Singhal
KIET Group of Institutions
Ghaziabad, India

Satyam Tomar
National Institute of Technology
Kurukshetra, India

Tarun Kumar Vashishth
IIMT University
Meerut, India

Navneet Verma
Panipat Institute of Engineering and
    Technology (PIET)
Samalkha, India

Ramesh Vishnoi
National Institute of Technology
Kurukshetra, India

Section 1

IoT and Technology

1 IoT-Based Work Monitoring
The Future of Productivity

T. S. Pradeep Kumar, Abhishek R. A.,
Renuka Devi Rajagopal, and Kumar Rangasamy

1.1 INTRODUCTION AND LITERATURE SURVEY

The use of technology in monitoring work habits and productivity has become increasingly popular in recent years. One such technology is the work monitoring system using IoT technology, which enables employers and parents to track and monitor the productivity and performance of their employees and children in real time. This system collects data continuously and provides real-time feedback to users, allowing for informed decisions and adjustments based on current work habits and productivity levels. The collected data can be analysed to gain insights into the work habits of individuals or teams, identifying areas of strength and potential areas for improvement. This information can be used to optimize workflow, improve efficiency, and identify problem areas that may be hindering productivity. The system can also be used to monitor and set performance goals, allowing users to track their progress and make adjustments as needed to stay on track.

Piyare and Tazil (2011) proposed a home automation system that is designed to be low-cost and flexible. The system is composed of two main components: a cell phone and an Arduino BT board. The system allows for the control of various home appliances by connecting them to the input/output ports of the Arduino BT board via relays. Kommey (2022) proposed an automatic fan regulation solution. To accomplish this, a temperature sensor and a camera were used with an intelligent system that tells whether the room is occupied or not. Soliman et al. (2017) suggested a smart home automation system that utilizes IoT technology. The proposed system has two options: a wireless and a wired scenario. Khoa et al. (2020) explored the leverage of using secure big data with cloud computing and IoT. The authors proposed Secure Hash Algorithm 256 (SHA-256) as an authentication mechanism to enhance security efficiency. Shouran et al. (2019) proposed smart homes that have various features such as real-time monitoring, remote control, intrusion prevention, gas/fire alarm, and more. Hoque and Davidson (2019) proposed a low-cost smart door sensor that alerts the status of the door to the users of either home or office. The work is integrated with Android OS along with RESTful API that uses Elegoo Mega 2560 microcontroller board along with Raspberry Pi2. Ektesabi et al. (2018) proposed an alternative smart

DOI: 10.1201/9781032673479-2

fan that addresses both comfort and cost considerations. To achieve this, the project employs a budget-friendly approach by utilizing a combination of readily available components from the market.

Rosli et al. suggested a smart home system that utilizes mobile app control through the IoT to educate Malaysians. It employs wireless technology, infrared sensors, and a microcontroller to monitor and manage the usage of household appliances such as LEDs, fans, and cameras. Rajesh et al. (2017) suggested a proposal for a smart home automation (HA) system that utilizes mobile technology to control various electric appliances based on signals sent by a mobile application. Balaji et al. (2020) proposed a project that aims to create a HA system specifically designed for kitchen wardrobes. The objective of the project is to develop a smart kitchen wardrobe that can automatically manage the grocery products stored in the kitchen wardrobe using wireless networks, smartphones, and the Internet. Goh et al. (2011) proposed a novel system for improved management of clothing collections by considering user preferences for colours, styles, events, and emotions. Specifically designed for busy entrepreneurs and individuals with colour blindness, the system integrates radio frequency identification technology into clothing items and tracks their movements within a wardrobe. Banoth et al. (2022) suggested the development of a smart wardrobe that notifies users about the availability of clothes or accessories in the wardrobe. The system employs sensing technology to capture images of the items and recognize them through a list.

Bagwari et al. (2019) suggested creating an Internet of Things (IoT) system that would enable users to access their wardrobe information from anywhere. The proposed system offers users the option to select clothes based on either cloud-stored events or current outdoor temperatures (Piyare and Tazil, 2011). This paper introduces a low-cost and flexible HA system that utilizes a cell phone and an Arduino BT board. The system is designed to connect home appliances to the input/output ports of the Arduino BT board through relays. Muthukrishnan et al. (2021) proposed a comparison of the features of Wi-Fi and LoRa for communication in agriculture. The IoT involves connecting devices to the Internet that work automatically without human intervention. Various wireless technologies support communication in IoT, including Wi-Fi, Bluetooth, ZigBee, LoRa, and NB-IoT. Talluri et al. (2020) proposed an automated and innovative water-showering mechanism that uses a rope-motor control system. This mechanism facilitates even distribution of water, fertilizers, and pesticides across a specific or entire area, thereby reducing unnecessary wastage. By using IoT technology for remote monitoring and control, we increase operational flexibility.

Pompigna and Mauro (2022) provide an overview of the smart road engineering approach by discussing the current state of innovation in the smart roads field. The goal is to improve energy efficiency and promote social, economic, and environmental sustainability. Balasubramaniam (2020) describes a proposed research project that aims to explore the main characteristics of ECG telemetry and develop an IoT-based application for monitoring patient health in both indoor and outdoor environments. Hu et al. (2018) proposed the architecture for home health care and monitoring, known as ICE (IoT Cares for Elderly), built on the Intel Edison platform. The system incorporates sensors that measure vital signs, sleeping patterns, and movements of humans. Park et al.'s (2017) study focuses on the challenges related to ambient assisted living and developing a system for the early detection of strokes. Ambient assisted living refers to the use of technology to support elderly and disabled individuals to live independently in their homes. Vishwakarma et al. (2019) propose a smart HA system that is energy-efficient and can control home devices from any location. An Internet connectivity module is attached to the main supply unit, allowing access through the Internet using a static IP address for wireless connectivity. The primary aim of this research is to improve the security and intelligence of the HA system.

Sokullu et al. (2020) proposed an intelligent home IoT system designed for elderly individuals and those suffering from memory impairment conditions such as MCI or dementia, which aims to provide safety and early warning mechanisms to assist them in coping with daily challenges associated with these disorders. Hoque and Davidson (2019) suggested an architecture for a smart door sensor that is designed to be cost-efficient and easy to implement. The sensor is intended to be used in homes and offices, and it alerts users through an Android app whenever a door is opened. Zhang et al. (2019) suggested the design of an intelligent device that automatically closes windows based on real-time weather conditions. The device can sense external information, analyse it, and convert it into an electric signal to control the mechanical part and complete the corresponding action. Aryal et al. (2018) proposed a vision and ongoing work of creating a smart IoT desk that can enhance the work environment around the occupant.

The desk is designed to act as a support system to drive the occupant's behaviour towards better environmental settings and to improve their posture and ergonomics. The smart IoT desk utilizes various sensors and intelligent algorithms to personalize the environment according to the occupant's preferences. For instance, the sensors can detect the ambient light, temperature, and noise levels in the room and adjust them accordingly. The desk can also remind the occupant to take breaks, sit up straight, and adjust their posture to avoid strain and injury. The primary goal of the smart IoT desk is to enhance the occupant's productivity and well-being by providing a personalized, comfortable, and ergonomic environment. The desk aims to make the work environment more pleasant and conducive to productive work by utilizing intelligent algorithms and various sensors. In conclusion, the paper presents the ongoing work of creating a smart IoT desk that can personalize the work environment and improve the occupant's posture and ergonomics. The desk aims to enhance the occupant's well-being and productivity by providing a comfortable, personalized, and ergonomic work environment.

In this chapter,

- Section 1.1 introduces the concept of work monitoring tools and their literature survey.
- Section 1.2 shows the system design and the problem statement for the work monitoring system.
- Section 1.3 specifies the hardware components involved in this work.
- Section 1.4 shows the experimental results and analysis of the work monitoring system using IoT.
- Section 1.5 shows the conclusion and future scope of the proposed device.

1.2 PROBLEM STATEMENT AND SYSTEM DESIGN

1.2.1 GENERAL

Design is a significant part of the development of a project, and it represents the meaning of the model that is to be built. Software design is a process through which the requirements are translated into a representation of the software. Design is the place where quality is rendered. Design is the means to accurately translate customer requirements into finished products.

1.2.2 BASIC ARCHITECTURE DIAGRAM

Figure 1.1 depicts the basic architecture diagram of how the project works. It shows that when the user is available and the sensor detects a person in front of it, it will start the predefined works such as turning on light, fan, etc. It stores the data when the phone is connected to the cloud and runs in real time. When no person is detected, the system will be off and no data will be fed to the cloud.

1.2.3 DATA FLOW DIAGRAM

Figure 1.2 depicts the flow of data from each component; at first, the Passive Infrared Sensor (PIR) sensor and temperature sensor detect the basic details, and then it sends to the NodeMicroController Unit (MCU). The NodeMCU commands the relay modules to work accordingly. These data are stored in the cloud, and then the Raspberry Pi is activated when a phone is connected, and those log details are then stored in the Firebase; all these details are displayed in the app.

This data flow diagram is depicted to show how information flows between the functionalities provided in the module. The data goes through some of the

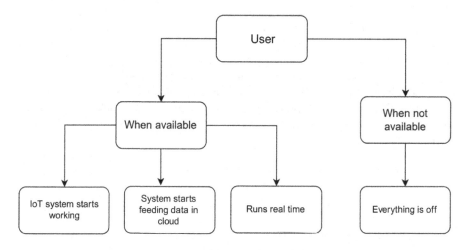

FIGURE 1.1   Basic architecture diagram.

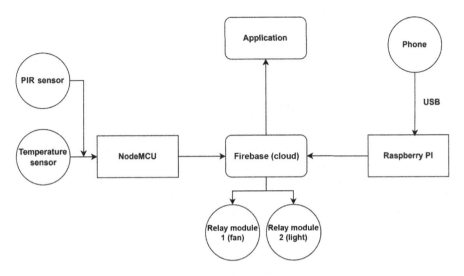

FIGURE 1.2 Data flow diagram.

predominant functionalities Extraction, Comparison, Calculation, and Detection. Incoming arrows represent the retrieval of data, and outgoing arrows represent inputting new data to storage.

1.2.4 Circuit Design

Figure 1.3 depicts the circuit diagram in which the circuit is connected, and it has a bulb and fan motor connected with relay modules.

1.3 HARDWARE COMPONENTS USED FOR IoT-BASED WORK MONITORING

The IoT-based system for monitoring student work utilizes several hardware components to function. These components work together to create an effective and user-friendly system.

1.3.1 NodeMCU ESP8266

The NodeMCU ESP8266 is a microcontroller board that serves as the main controller for the system. It is Wi-Fi-enabled, which means it can connect to the Internet and send data to the cloud. The NodeMCU ESP8266 is programmed using the Arduino IDE, which is an integrated development environment used to write and upload code to the board. It is designed to make it easy for developers to build Wi-Fi-enabled projects with the ESP8266, as it provides a high-level interface for programming and accessing the module's features. NodeMCU is programmed using Lua, a scripting

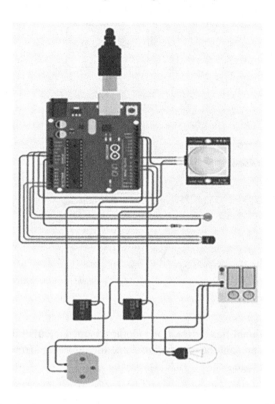

FIGURE 1.3   Circuit diagram.

FIGURE 1.4   NodeMCU (ESP 8266).

language that is lightweight and easy to programme. The firmware provides a set of modules that allow developers to interact with the ESP8266's GPIO pins, Wi-Fi functionality, file system, and other features. NodeMCU has become a popular platform for IoT projects due to its ease of use and low cost. It is compatible with a variety of sensors and I/O devices that can be programmed to connect to cloud services such as Amazon Web Services (AWS) and Google Cloud Platform (Figure 1.4).

1.3.2 RASPBERRY PI

Raspberry Pi is a small but powerful single-board computer developed at the UK by the RPi Foundation. Raspberry Pi boards are powered by ARM-based processors and run a variety of operating systems, including the Linux-based Raspberry Pi OS. They have a range of connectivity options, including Wi-Fi and Ethernet, and several GPIO pins that can be used to connect sensors, actuators, and other electronic components. Raspberry Pi has a large and active community of developers and enthusiasts who have created a wide range of projects, from media centres and game consoles to weather stations and HA systems. The low cost and versatility of the Raspberry Pi make it an ideal platform for prototyping and experimenting with a wide range of projects (Figure 1.5).

1.3.3 DHT11 SENSOR

The DHT11 is a low-cost digital temperature and humidity sensor that can be used in a wide range of applications. It is a basic sensor that provides accurate readings of temperature and humidity with a resolution of 1°C and 1% relative humidity (Figure 1.6).

The DHT11 sensor is a single-wire digital sensor, meaning that it requires only one data pin to communicate with a microcontroller. The sensor has a measuring range of 0°C–50°C for temperature and 20%–90% for relative humidity, with an accuracy of ±2°C and ±5% relative humidity.

1.3.4 PIR SENSOR

A PIR (passive infrared) sensor is a type of motion sensor that detects the presence of people or animals by detecting changes in infrared radiation in its field of view. The sensor works by detecting changes in the infrared energy emitted by objects in its field of view, which are caused by their movement. PIR sensors are commonly used in

FIGURE 1.5 Raspberry Pi4.

FIGURE 1.6   DHT11 sensor (temperature and humidity).

FIGURE 1.7   Passive infrared sensor (PIR sensor).

security systems, lighting control systems, and other applications where detecting the presence of people or animals is important. They are easy to use and install, and can be connected to microcontrollers or other devices using digital or analogue inputs. They are typically powered by batteries or a low-voltage power supply and consume very little power, making them ideal for battery-powered applications (Figure 1.7).

1.3.5   Relay Module

A relay module is an electronic device that allows a low-power signal to control a high-power load. It consists of one or more relays, which are electromechanical switches that can be controlled by a microcontroller or other low-power device. The relay module provides isolation between the control circuit and the load, allowing them to operate at different voltages or currents. Relay modules are commonly used

FIGURE 1.8 Relay module.

in industrial automation, HA, and other applications where it is necessary to control high-power loads such as motors, lights, or appliances. They are available in a range of sizes and configurations, from single-channel to multi-channel modules, and can be controlled by a variety of inputs including digital signals, analogue signals, or even wireless signals. Relay modules are easy to use and provide a reliable and flexible way to control high-power loads using a low-power signal. They are an important component in many electronic and automation systems, and are widely available at affordable prices (Figure 1.8).

1.3.6 FLUTTER

Flutter is a mobile application development platform framed by Google for the cross-applications and portability to various mobile apps namely Android and iOS. This API contains a single codebase that uses Dart as a programming language, which is a fast and modern language that provides features such as hot reloading, which allows for quick iteration during development. One of the key advantages of Flutter is its ability to create visually attractive and customizable user interfaces, thanks to its built-in widget library. Flutter widgets are designed to be highly customizable and can be combined to create complex user interfaces with animations and effects. Flutter also provides a rich set of tools and libraries for building apps, including plugins for accessing device sensors and services, network connectivity, and database storage. This makes it easy to create full-featured and responsive mobile apps with a consistent user experience across both platforms. Flutter has gained popularity among developers due to its ease of use, flexibility, and performance. It has been used to develop apps for a wide range of industries, including finance, health, education, and entertainment. The framework is actively developed and maintained by Google and the Flutter community, ensuring that it remains up-to-date with the latest technologies and trends in mobile development.

1.3.7 FIREBASE

Firebase Cloud is a cloud-based platform provided by Google for developing and hosting mobile and web applications. Firebase Cloud offers a wide range of services including authentication, cloud storage, real-time database, hosting, and functions. These services can be used individually or combined to create powerful and scalable applications. Firebase Cloud services are built on Google's cloud infrastructure, which provides a high level of reliability and scalability. This ensures that applications built on Firebase Cloud can handle high traffic loads and remain available even during peak usage periods. Firebase Cloud offers several benefits to developers, including easy integration with popular mobile and web development frameworks such as Flutter and React Native.

It also provides a simple and intuitive interface for managing applications and configuring services. Firebase Cloud also offers real-time synchronization of data across devices, which makes it easy to create collaborative and interactive applications. This is made possible by the real-time database service, which allows developers to store and retrieve data in real time. Firebase Cloud offers a pay-as-you-go pricing model, which makes it affordable for developers and businesses of all sizes. Additionally, Firebase Cloud provides extensive documentation and support, making it easy for developers to get started with the platform and troubleshoot any issues that may arise. Overall, Firebase Cloud is a powerful and flexible platform for developing and hosting mobile and web applications. Its wide range of services and ease of use make it a popular choice among developers looking to build scalable and reliable applications.

1.4 EXPERIMENTAL EVALUATION AND RESULTS

1.4.1 RESULTS AND CONCLUSION

The project is ready as per the above-mentioned proposed system. It may have a huge impact on the lives of students and organizations, and it not only has the features mentioned above but also has energy-saving benefits, when people are not around all the electronic devices turns off. The images of the outputs are attached below.

1.4.1.1 Firebase Cloud

All the data syncing is done here, and here is where the app gets the details from.

Figure 1.9 is a visual representation of the connection of mobile phones. The figure shows various elements such as the names of the connected devices, their log details, and the overall connectivity between them.

Figure 1.10 is a visual representation of the disconnection details of a mobile phone. The figure shows various elements such as the log details, the device name, and the time of disconnection.

Figure 1.11 depicts the email id and name by which it is logged in and the password is hidden.

Figure 1.12 depicts that the phone number is also shown in the cloud. Users can add up to two phone numbers while signing in.

1.4.1.2 Environmental Control System

The diagram in Figure 1.13 displays the results of the environmental control system segment of the project. It reveals that when the user is in front of the system and the

FIGURE 1.9 Database output.

FIGURE 1.10 Database output.

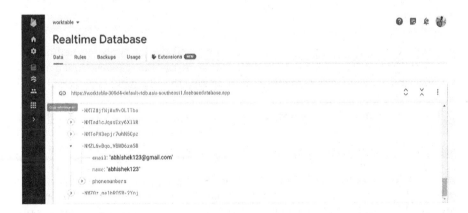

FIGURE 1.11 Database output.

FIGURE 1.12   Database output.

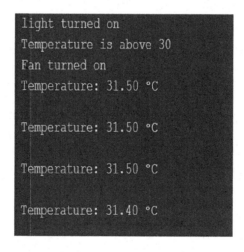

FIGURE 1.13   Environment control system.

temperature exceeds 30°, the light is switched on, and the fan is also turned on. In other words, the image illustrates how the system operates in response to the user's presence and the temperature condition, illuminating the area with the light and providing cooling through the fan.

The diagram in Figure 1.14 displays the results of the environmental control system segment of the project. It reveals that when the user is in front of the system and the temperature exceeds 30°, the light is switched on, and the fan is also turned on.

Figure 1.15 demonstrates that the person has exited the room, and the passive infrared (PIR) sensor installed in the system is no longer detecting any presence in front of it. Therefore, the system has responded by turning off all the connected devices. This could include the lights, fan, and any other appliances that are parts of

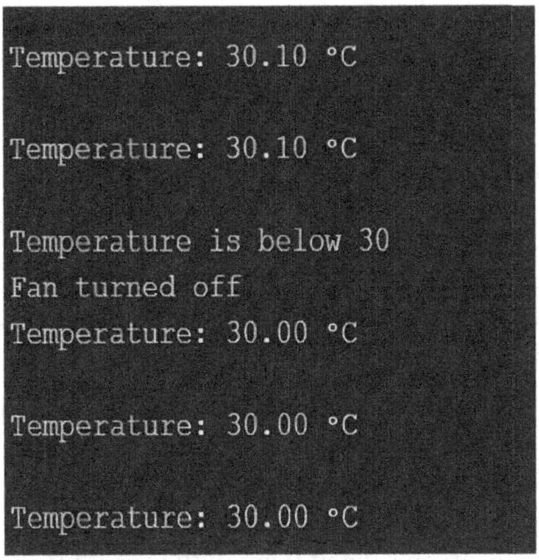

```
Temperature: 30.10 °C

Temperature: 30.10 °C

Temperature is below 30
Fan turned off
Temperature: 30.00 °C

Temperature: 30.00 °C

Temperature: 30.00 °C
```

FIGURE 1.14 Environment control system.

```
Temperature: 29.70 °C

Temperature: 29.70 °C

Temperature: 29.70 °C

Temperature: 29.70 °C

Person left
light turned off
```

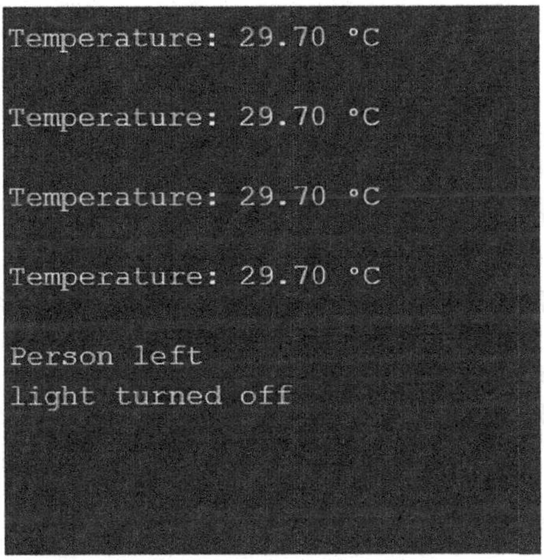

FIGURE 1.15 Environment control system.

the control system. The purpose of this action is to conserve energy and avoid wasting electricity when there is no need for it.

1.5 CONCLUSION AND FUTURE WORK

The evaluation of the proposed smart desk with a work monitoring system was concluded through an experimental study involving five participants from diverse backgrounds. The study was conducted over a period of 4 weeks, during which the participants used the system to monitor their work habits and productivity.

The results of the study showed that the smart desk with a work monitoring system was effective in improving the participants' work habits and productivity. Specifically, the following results were observed:

- **Improved Time Management**: Participants reported that the system helped them manage their time more effectively, by providing them with reminders and notifications to stay on track.
- **Increased Productivity**: Participants reported that the system helped them increase their productivity, by providing them with real-time feedback on their work habits and progress.
- **Enhanced Accountability**: Participants reported that the system helped them become more accountable for their work, by providing them with data on their work habits and progress.
- **Positive User Experience**: Participants reported that the system was easy to use and navigate, and provided them with useful information to optimize their work habits.

Overall, the results of the study suggest that the proposed smart desk with a work monitoring system has the potential to improve work habits and productivity, making it a valuable tool for individuals and organizations.

In order to further improve the product, software updates can be implemented, the task manager can be enhanced, and various other features can be added. Additionally, phone detection software can be developed for iPhone Operating System (IOS). Moreover, the smart study table can incorporate accessibility features such as text-to-speech and speech-to-text capabilities, enabling students with disabilities to participate in classroom learning. Furthermore, artificial intelligence and machine learning algorithms can be utilized by the smart study table to tailor the learning experience according to the student's individual strengths and weaknesses. This can include recommending learning materials and activities specific to each student's learning style and pace. Lastly, the smart study table can integrate virtual reality technology, providing students with an immersive and interactive learning experience. For instance, students can use virtual reality headsets to explore historical sites or scientific phenomena, enhancing engagement and promoting better retention.

REFERENCES

Aryal, A., Anselmo, F., & Becerik-Gerber, B. (2018). Smart IoT desk for personalizing indoor environmental conditions. In *Proceedings of the 8th International Conference on the Internet of Things* (pp. 1–6), Santa Barbara, California, USA.

Bagwari, S., Singh, R., & Gehlot, A. (2019). Internet of things based intelligent wardrobe. In *International Conference on Advances in Engineering Science Management & Technology (ICAESMT)-2019*, Uttaranchal University, Dehradun, India.

Balaji, A., Sathyasri, B., Vanaja, S., Manasa, M. N., Malavega, M., & Maheswari, S. (2020). Smart kitchen wardrobe system based on IoT. In *2020 International Conference on Smart Electronics and Communication (ICOSEC)* (pp. 865–871), Trichy, India. IEEE.

Balasubramaniam, V. (2020). IoT based biotelemetry for smart health care monitoring system. *Journal of Information Technology and Digital World, 2*(3), 183–190.

Banoth, R., Godishala, A., Yassin, H., & Veena, R. (2022). Next generation smart wardrobe management system using IoT. In *2022 IEEE 7th International Conference for Convergence in Technology (I2CT)* (pp. 1–4), Pune, India. IEEE.

Ektesabi, M., Gorji, A. S., Moradi, A., Yammen, S., Reddy, V. M., & Tang, S. (2018). IOT-based home appliance system (SMART FAN). *Computer Science & Information Technology (CS & IT), 8*(16), 37–46.

Goh, K. N., Chen, Y. Y., & Lin, E. S. (2011). Developing a smart wardrobe system. In *2011 IEEE Consumer Communications and Networking Conference (CCNC)* (pp. 303–307). IEEE.

Hoque, M. A., & Davidson, C. (2019). Design and implementation of an IoT-based smart home security system. *International Journal of Networked and Distributed Computing, 7*(2), 85–92.

Hu, B. D. C., Fahmi, H., Yuhao, L., Kiong, C. C., & Harun, A. (2018). Internet of Things (IOT) monitoring system for elderly. In *2018 International Conference on Intelligent and Advanced System (ICIAS)* (pp. 1–6), Kuala Lumpur, Malaysia. IEEE.

Khoa, T. A., Nhu, L. M. B., Son, H. H., Trong, N. M., Phuc, C. H., Phuong, N. T. H., ... Duc, D. N. M. (2020). Designing efficient smart home management with IoT smart lighting: A case study. *Wireless Communications and Mobile Computing, 2020*(1), 8896637.

Kommey, B. (2022). Automatic ceiling fan control using temperature and room occupancy. *Journal of Information Technology and Computer Engineering, 6*(01), 1–7.

Muthukrishnan, H., Jeevanantham, A., Sunita, B., Najeerabanu, S., & Yasuvanth, V. (2021). Performance analysis of Wi-Fi and LoRa technology and its implementation in farm monitoring system. *IOP Conference Series: Materials Science and Engineering, 1055*(1), 012051.

Park, S. J., Subramaniyam, M., Kim, S. E., Hong, S., Lee, J. H., Jo, C. M., & Seo, Y. (2017). Development of the elderly healthcare monitoring system with IoT. In *Advances in Human Factors and Ergonomics in Healthcare: Proceedings of the AHFE 2016 International Conference on Human Factors and Ergonomics in Healthcare*, July 27–31, 2016, Walt Disney World(r), Florida, USA (pp. 309–315). Springer International Publishing.

Piyare, R., & Tazil, M. (2011). Bluetooth-based home automation system using a cell phone. In *2011 IEEE 15th International Symposium on Consumer Electronics (ISCE)* (pp. 192–195), Singapore. IEEE.

Pompigna, A., & Mauro, R. (2022). Smart roads: A state of the art of highways innovations in the smart age. *Engineering Science and Technology, an International Journal, 25*, 100986.

Rajesh, T., Rahul, R., Malligarjun, M., & Suvathi, M. (2017). Home automation using a smartphone application. *International Journal of Advanced Research in Science Engineering and Technology, 4*(3), 3546–3553.

Rosli, A. A., Razif, M. R. M., Hassan, O. A., Shah, N. S. M., Nordin, I. N. A. M., & Mustafa, K. N. (2021) Development of Smart Home System based on Mobile Apps Control Using IoT for Educational Purposes. *Alinteri J. of Agr. Sci. 36*(1): 525–533

Shouran, Z., Ashari, A., & Priyambodo, T. (2019). Internet of things (IoT) of smart home: Privacy and security. *International Journal of Computer Applications, 182*(39), 3–8.

Sokullu, R., Akkaş, M. A., & Demir, E. (2020). IoT-supported smart home for the elderly. *Internet of Things, 11*, 100239.

Soliman, M. S., Dwairi, M. O., Sulayman, I. I. A., & Almalki, S. H. (2017). Towards the design and implementation, of a smart home automation system based on the Internet of Things approach. *International Journal of Applied Engineering Research, 12*(11), 2731–2737.

Talluri, M. T., Arya, S. S., Tripathi, A., & Karthikeyan, V. (2020). Iot-based multipurpose uniform water showering mechanism for urban agriculture. In *2020 IEEE First International Conference on Smart Technologies for Power, Energy, and Control (STPEC)* (pp. 1–6), Nagpur, India. IEEE.

Vishwakarma, S. K., Upadhyaya, P., Kumari, B., & Mishra, A. K. (2019). Smart energy efficient home automation system using IoT. In *2019 4th International Conference on Internet of Things: Smart Innovation and Usages (IoT-SIU)* (pp. 1–4), Ghaziabad, India. IEEE.

Zhang, K., Shi, G., & Zhai, Z. (2019). Design and Implementation of automatic window closer based on intelligent control algorithm. In *2019 IEEE International Conference on Mechatronics and Automation (ICMA)* (pp. 566–571), Tianjin, China. IEEE.

2 Harnessing the Power of IoT and Big Data

Advancements and Applications in Smart Environments

*Anita Mohanty, Ambarish G. Mohapatra,
Subrat Kumar Mohanty, and Sasmita Nayak*

2.1 INTRODUCTION

The convergence of Internet of Things (IoT) and Big Data technologies marks a pivotal and momentous juncture in the digital landscape, setting ablaze a profound and far-reaching transformation across diverse industries and domains. At the very core of this evolution lies the seamless integration of IoT devices, forging connections between ordinary objects and systems, and the formidable analytical prowess of Big Data, diligently processing and extracting invaluable insights from vast and diverse datasets [1]. This formidable amalgamation unfurls a boundless realm of possibilities, paving the way for the unprecedented development of smart environments that possess the potential to revolutionize our lifestyles, work dynamics, and the very fabric of our interactions with the world.

2.1.1 THE SIGNIFICANCE OF IoT AND BIG DATA CONVERGENCE

The profound synergy brought forth by the convergence of IoT and Big Data is instigating a paradigm shift in the acquisition, processing, and utilization of information. IoT devices, encompassing an array of sensors, wearable, and interconnected machines, orchestrate the generation of an unyielding torrent of real-time data, offering an uninterrupted flow of insights from the physical world. On the other hand, the formidable might of Big Data analytics rises to the occasion, diligently unraveling intricate patterns, illuminating hidden trends, and revealing profound correlations hitherto concealed [2]. This transformative convergence empowers organizations and individuals alike to make informed and data-driven decisions, derive actionable and prescient insights, and optimize operational frameworks like never before. The capability to assimilate, scrutinize, and respond to real-time data fortifies an array of

DOI: 10.1201/9781032673479-3

19

industries, extending from healthcare and transportation to energy and urban planning, instilling them with a newfound prowess to embrace efficiency, resilience, and
adaptability in the face of dynamic and ever-changing demands.

2.1.2 AN OVERVIEW OF SMART ENVIRONMENTS AND THEIR BOUNTIFUL ADVANTAGES

Smart environments, the pinnacle of IoT and Big Data applications, manifest as
enthralling and interconnected landscapes where devices and intelligent systems
harmoniously collaborate to embellish our surroundings with unprecedented effi-
ciencies and enhancements [3]. These futuristic environments find diverse incar-
nations in smart homes, smart cities, smart factories, and even smart healthcare
facilities, as shown in Figure 2.1. Within smart homes, IoT devices dutifully shoul-
der the mantle of automating tasks, optimizing energy consumption, and fortifying
security measures, offering the occupants an immensely convenient and uniquely
personalized living experience. In the realm of smart cities, IoT sensors intertwine
with the wizardry of data analytics to herald an era of improved traffic management,
resource allocation optimization, and fortified public services, fostering sustainability and elevating the quality of life for citizens at large. Simultaneously, across
industries such as manufacturing and healthcare, the realm of smart environments
streamlines operations, curtails downtime, and beckons forth the era of predictive
maintenance, leading to heightened productivity and delightful cost savings. The
seemingly boundless potential benefits of smart environments proffer an array of

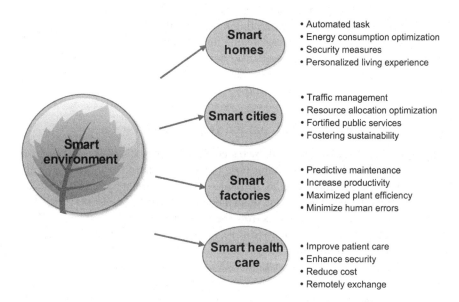

FIGURE 2.1   Smart environment: an overview.

solutions to the myriad challenges confronting societies and industries as they diligently chart the path toward an interconnected, enlightened, and intelligent future.

As we venture deep into the enthralling domain of IoT and Big Data applications within various smart environments, a panorama laden with opportunities and innovations unfurls before our very eyes [4]. However, amid these exciting opportunities lies the imperative to grapple with equally consequential challenges, including those concerning data privacy, security, and scalability. The concerted endeavor to address and surmount these formidable obstacles is of paramount significance, for it holds the key to realizing the full extent of the transformative power embedded within the convergence of IoT and Big Data, bestowing upon us the unfettered capacity to usher in a new era of smart environments, poised to redefine the boundaries of human potential. Within the confines of this chapter, we embark on an odyssey of exploration, venturing deep into the intricate interplay of IoT and Big Data, and illuminating the path that leads us to the vanguard of innovation, as we shape and mold the intelligent ecosystems of tomorrow, with unprecedented insights into the groundbreaking applications that shall inexorably shape the world that lies ahead.

The chapter is organized into several sections that explore the convergence of IoT and Big Data in the context of smart environments. It begins with an introduction, highlighting the significance of this convergence and the potential benefits it offers. The subsequent sections delve into specific domains of smart environments, starting with smart transportation, which covers IoT-powered traffic management and real-time analytics. Smart Healthcare follows, discussing remote monitoring, wearable sensors, and Big Data's role in personalized treatment. Smart Grid is the next focus, examining applications in energy management and grid stability. A Smart Inventory System is explored with its supply chain optimization and inventory management through IoT. The chapter then delves into smart city, touching on infrastructure, waste management, and public safety. Section 2.7 addresses security, scalability, and the future of IoT-driven smart environments. The chapter concludes by summarizing key findings, discussing implications, and envisioning a connected and data-driven future.

2.2 SMART TRANSPORTATION

Smart transportation refers to the integration of cutting-edge technology and innovative solutions into the traditional transportation infrastructure to create more efficient, sustainable, and convenient mobility options. It encompasses a wide range of advancements such as autonomous vehicles, intelligent traffic management systems, real-time data analysis, and seamless connectivity between various modes of transportation [5]. Smart transportation aims to reduce traffic congestion, lower emissions, enhance safety, and improve the overall quality of transportation services, as mentioned in Figure 2.2. By influencing data-driven insights and automation, it commits to revolutionizing the way people and goods move within urban and rural environments, ultimately shaping a more related and environmentally friendly future for transportation.

2.2.1 IoT-POWERED INTELLIGENT TRAFFIC MANAGEMENT

The burgeoning phenomenon of urbanization comes with a multitude of complex challenges, and among them, traffic congestion stands as a pressing concern.

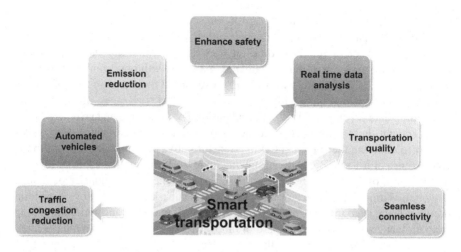

FIGURE 2.2   Advancement in smart transportation system.

As cities expand, traditional traffic management systems grapple with the ever-increasing demands of urban mobility. Enter IoT-powered intelligent traffic management, a pioneering solution harnessing the capabilities of the IoT to redefine our approach to traffic control and urban transportation [6,7].

At its core, this system revolves around the strategic deployment of interconnected sensors, cameras, and data analytics tools throughout urban infrastructure. These elements work in tandem to collect and transmit real-time data on traffic conditions, road usage, and environmental variables. This dynamic repository of real-time data empowers city planners and traffic authorities with comprehensive insights, facilitating data-driven decision-making and swift responses to fluctuating traffic dynamics and incidents.

One of the standout features of this system is its capacity for dynamic traffic control. Intelligent traffic lights, for instance, adapt signal timings based on real-time traffic data, curtailing unnecessary stops, mitigating congestion, and curbing fuel consumption and emissions.

Furthermore, IoT-driven predictive analytics transcend real-time data, forecasting traffic patterns and potential incidents, enabling commuters to plan routes wisely and authorities to proactively manage traffic. Safety receives a substantial boost through prompt accident detection by IoT sensors, facilitating automatic alerts to emergency services, thereby reducing response times and potentially saving lives.

Sustainability is a focal point, as IoT-powered traffic management optimizes traffic flow, diminishes congestion, and encourages eco-friendly transportation modes, ultimately reducing carbon emissions and enhancing air quality. Efficiency, the core tenet, is manifested through the reduction of delays, minimized idling, and heightened overall transportation system performance, translating into time, resource, and energy savings (Figure 2.3) [8].

The treasure trove of data amassed by IoT devices empowers city planners in making informed decisions concerning infrastructure enhancements, traffic management

FIGURE 2.3 IoT-driven intelligent traffic management system.

strategies, and safety protocols. Everyday commuters are not left behind, benefiting from real-time traffic updates, which enable informed choices on alternative routes or transportation modes, alleviating the vexations associated with traffic congestion.

2.2.2 Real-Time Data Analytics for Traffic Prediction and Optimization

In the ever-expanding metropolitan landscapes of the 21st century, traffic jam has become a universal challenge, impacting everything from daily commutes to economic productivity and environmental sustainability. However, there's an alarm of hope on the horizon – real-time data analytics for traffic forecast and optimization [9,10]. This cutting-edge technology is adapting how cities manage their transportation networks and gives a glimpse into a prospective of smoother, more efficient, and eco-friendly urban mobility.

- **Real-Time Data Gathering**: The base of this transformative approach lies in the deployment of sensors, cameras, and interconnected equipments throughout city infrastructure. These devices repeatedly gather real-time data on traffic flow, road conditions, and even weather patterns.
- **Immediate Insights**: The data gathered is processed in real time, providing city authorities and transportation managers with instantaneous insights into the state of the traffic network. This allows swift decision-making and proactive replies to changing conditions.
- **Dynamic Traffic Control**: Real-time data analytics allow traffic management systems to dynamically manage traffic flow. For instance, traffic lights can adopt their timings on the fly based on actual traffic volumes, decreasing unnecessary stops, minimizing congestion, and reducing fuel consumption and emissions.
- **Predictive Analytics**: Beyond just reacting to current conditions, these systems use historical data and machine learning algorithms to forecast future traffic patterns. Commuters and city planners can gain from this foresight, making more informed choices and optimizing routes and schedules.
- **Enhanced Safety**: Real-time analytics play a vital role in enhancing road safety. They can swiftly detect accidents or road hazards, alerting emergency services and allowing for quicker response times, thereby potentially saving lives.
- **Sustainability**: Traffic optimization through real-time data analytics contributes significantly to sustainability efforts. Reducing traffic congestion and encouraging the use of public transport or carpooling helps lower greenhouse gas emissions and improve air quality.
- **Efficiency**: Efficiency is at the heart of this technology. By minimizing delays, reducing idling time, and enhancing overall transportation system performance, cities save valuable time, resources, and energy.
- **Data-Driven Decision-Making**: The wealth of data collected aids city planners and transportation authorities in making data-driven decisions. It informs infrastructure improvements, traffic management strategies, and safety measures.
- **Empowering Commuters**: Everyday, commuters become active participants in the optimization process. They receive real-time traffic updates through navigation apps, allowing them to make on-the-spot decisions to avoid traffic bottlenecks and reach their destinations more efficiently.
- **Building Smart Cities**: In conclusion, real-time data analytics for traffic prediction and optimization is a transformative force that is shaping the cities of tomorrow. As urbanization continues its relentless march, this technology ensures that transportation networks evolve to meet the demands of the modern world, fostering smarter, safer, and more sustainable urban environments for all to enjoy. It is a beacon of hope that illuminates the path to a brighter, more efficient, and interconnected future.

2.2.3 Connected Vehicles and V2I/V2V Communication

Connected vehicles, coupled with Vehicle-to-Infrastructure (V2I) and Vehicle-to-Vehicle (V2V) communication, represent a pivotal advancement in the automotive industry and the future of transportation [11]. These smart vehicles are equipped with sensors, cameras, and communication systems that enable them to exchange data not only with each other but also with traffic infrastructure such as traffic lights, road signs, and even pedestrians' smartphones. V2I communication permits vehicles to collect real-time traffic information, optimize routes, and enhance safety by interacting with traffic management systems. On the other hand, V2V communication empowers vehicles to share critical information like speed, location, and potential hazards with nearby vehicles, assisting the progress of collision avoidance and creating a safer driving environment [12]. Together, these technologies agree to reduce traffic congestion, raise road safety, and pave the way for more efficient and interconnected transportation systems in our increasingly urbanized world.

2.2.4 Success Stories of IoT-Driven Smart Transportation Initiatives

The integration of the IoT into transportation systems has start numerous success stories worldwide, demonstrating the transformative power of technology in making urban mobility more effective, sustainable, and convenient. Here are a few important examples of IoT-driven smart transportation initiatives that have made an important impact:

- **Singapore's Smart Traffic Management:** Singapore, known for its forward-thinking access, has fitted an extensive IoT-powered smart traffic management system [13]. The city-state handles sensors and cameras to monitor traffic conditions in real time. This data feeds into an intelligent traffic control system that adapts traffic signals, manages congestion, and gives real-time traffic updates to commuters. Singapore's smart transportation actions have significantly decreased congestion and enhanced travel times.
- **Barcelona's Smart Parking:** Barcelona's IoT-driven smart parking system is an alarm of efficiency in city transportation. Using sensors embedded in parking spaces, the city gives real-time information on free parking spots through mobile apps and digital signage. This not only decreases the time drivers spend searching for parking but also lowers emissions by minimizing irrelevant circling in search of a spot [14].
- **London's Connected Buses:** With IoT connectivity, London's classic red buses have gone modern. These buses have sensors that gather information on traffic, weather, and passenger counts. Using this information, real-time bus route optimization ensures that buses arrive when and where they are most needed, cutting down on waiting times and improving the appeal of public transit.
- **Los Angeles' Automated Traffic Management:** In order to predict traffic congestion and modify signal timings accordingly, Los Angeles has built

an IoT-driven Automated Traffic Management System that makes use of sensors, cameras, and machine learning algorithms [15]. The overall traffic flow has improved and delays caused by traffic have been significantly reduced thanks to this dynamic strategy.

- **Amsterdam's Smart Canal Boats:** IoT advancements can even help with maritime shipping. Smart canal boats have been introduced in Amsterdam that employ IoT sensors to keep an eye on weather, traffic density, and water levels [16]. This information aids in streamlining canal traffic, guaranteeing efficient and secure travel for locals and visitors alike.
- **India's Toll Collection System:** The National Highways Authority of India has put in place FASTag, an IoT-based electronic toll collection system [17]. It makes use of radio-frequency identification (RFID) technology to enable cashless, contactless toll payments, easing traffic congestion at toll booths and enhancing overall highway traffic flow.
- **Sydney's Smart Traffic Lights:** IoT-connected traffic signals are assisting in easing congestion and enhancing traffic flow in Sydney, Australia. The city's road network is made more effective by these lights, which employ real-time data to modify signal timings based on traffic circumstances, cutting commuter trip times.

2.3   SMART HEALTHCARE

Healthcare is changing drastically in the digital age, and the idea of "smart healthcare" is at the core of this transition. In order to improve patient care and healthcare delivery and make medical procedures more accessible and efficient, smart healthcare makes use of cutting-edge technologies, data analytics, and networking [18]. Figure 2.4 shows the outline of the medical scene that changes due to the implementation of smart healthcare system:

- **Remote Patient Monitoring:** Wearable technology and sensors are used in smart healthcare to enable ongoing monitoring of patients' vital signs and health metrics. Especially for chronic illnesses, this real-time data can be provided to healthcare providers, enabling proactive intervention, early disease detection, and individualized treatment strategies.
- **Telemedicine and Telehealth:** Patients have remote access to healthcare providers thanks to telemedicine platforms and mobile health apps. This not only improves access to healthcare, particularly in rural areas, but also lessens the demand on healthcare institutions, improving the effectiveness and convenience of healthcare services.
- **Electronic Health Records (EHRs):** Medical data is easily accessible to authorized healthcare professionals thanks to the centralization of patient data in TelemedicSmart healthcare systems' electronic health records. Informed decision-making is encouraged, medical errors are decreased, and administrative procedures are streamlined.

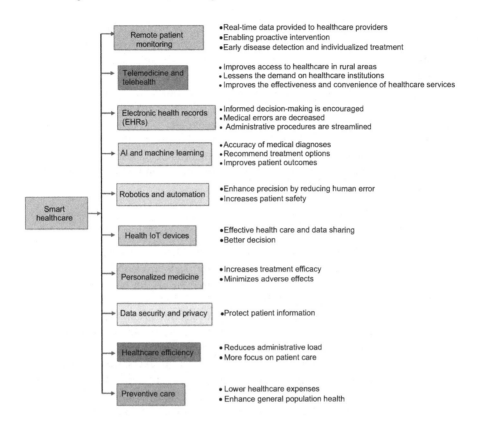

FIGURE 2.4 Outline of smart healthcare system.

- **AI and Machine Learning:** Artificial intelligence and machine learning algorithms analyze vast healthcare datasets to identify patterns, predict disease outbreaks, and assist in medical diagnostics. These technologies enhance the accuracy of medical diagnoses, recommend treatment options, and improve patient outcomes.
- **Robotics and Automation:** Smart healthcare incorporates robotics for tasks such as surgery, medication dispensing, and patient care. These robotic systems enhance precision and reduce the risk of human error, ultimately improving patient safety.
- **Health IoT Devices:** The IoT has enabled the proliferation of smart medical devices, from smart thermometers to pill dispensers. These devices help individuals manage their health more effectively and share data with healthcare professionals for better-informed decisions.
- **Personalized Medicine:** With the help of genetic profiling and data analytics, smart healthcare tailors treatment plans to individual patient's genetic makeup, lifestyle, and medical history. This precision medicine approach increases treatment efficacy and minimizes adverse effects.

- **Data Security and Privacy:** Smart healthcare prioritizes robust data security measures to protect patient information. Strict privacy regulations, such as HIPAA in the United States, ensure that healthcare data remains confidential and secure.
- **Healthcare Efficiency:** Automation and data-driven insights optimize healthcare workflows, reducing administrative overhead and allowing healthcare providers to focus more on patient care.
- **Preventive Care:** Through health monitoring, early intervention, and patient education, smart healthcare places a strong emphasis on preventive measures. This proactive strategy seeks to lower healthcare expenses and enhance general population health.

2.3.1 REMOTE HEALTH MONITORING USING IoT DEVICES

Remote health monitoring is a revolutionary method that has emerged as a result of the IoT convergence with the healthcare industry. This groundbreaking idea uses IoT gadgets with sensors and connectivity to keep an eye on patients' health and well-being from the convenience of their homes or almost anywhere else [19,20]. By strengthening patient care, lowering healthcare costs, and promoting overall health outcomes, remote health monitoring has the potential to change healthcare.

Continuous Patient Monitoring: Vital signs and health metrics can be continuously monitored using IoT devices like wearable fitness trackers, smartwatches, and medical-grade sensors. These gadgets can monitor data like temperature, blood glucose levels, heart rate, blood pressure, and even sleep habits. For those with chronic illnesses, this real-time data is essential since it enables medical professionals to closely monitor their symptoms without needing to see them frequently.

Timely Intervention: In the event of irregularities or emergencies, remote health monitoring enables prompt response. When predetermined criteria are exceeded, IoT devices can notify both patients and medical staff, enabling quick action and lowering the possibility of problems.

Chronic Disease Management: Remote monitoring offers a lifeline to people who are managing chronic diseases like diabetes, hypertension, or heart disease. In order to create more individualized treatment plans and enhance illness management, patients can regularly track their health indicators and share this information with their healthcare teams.

Medication Adherence: IoT devices can also help ensure medication adherence. Smart pill dispensers can remind patients to take their medications on schedule and send alerts to caregivers or healthcare providers if doses are missed.

Telehealth Integration: Remote health monitoring seamlessly integrates with telehealth and telemedicine platforms. Patients can share their health data with healthcare providers during virtual consultations, enabling more informed and productive discussions and reducing the need for physical office visits.

Improved Patient Engagement: Patients become more engaged in their healthcare when they have access to their health data. IoT devices empower individuals to take control of their well-being, make informed lifestyle choices, and actively participate in their treatment plans.

Reduced Healthcare Costs: By preventing hospital readmissions, minimizing emergency room visits, and facilitating early interventions, remote health monitoring has the potential to reduce healthcare costs significantly. It promotes a shift from reactive care to proactive, preventive healthcare.

Elderly Care: Remote monitoring is particularly valuable for elderly individuals who may have multiple chronic conditions and require ongoing care. Seniors can age in place and yet receive top-notch medical care thanks to it.

Data Security: Remote health monitoring systems abide by stringent healthcare rules, such as HIPAA in the United States, to guarantee patient data security and privacy. This guarantees that private health information is kept private and secure.

Scalability and Accessibility: Because remote health monitoring is scalable, it is available to a wide spectrum of patients, including those in underserved or rural locations as well as those in densely populated urban areas.

2.3.2 WEARABLE SENSORS AND HEALTH DATA INTEGRATION

Wearable sensors have permeated every aspect of our life; they now adorn our wrists as smartwatches, count our steps and heart rates when exercising, and even keep tabs on our sleeping habits. Through the integration of health data, these gadgets are ushering in a new era of healthcare in addition to their convenience and fitness-focused applications [21]. The way people manage their health is being revolutionized by this fusion of technology and healthcare, and it is also enabling healthcare professionals to provide more individualized and efficient care.

- **Holistic Health Tracking:** A comprehensive approach to health monitoring is provided by wearable sensors. They gather information about different health markers, such as heart rate, physical activity, sleep quality, and even stress levels, continually. This round-the-clock observation offers a whole picture of a person's health, enabling a more complex comprehension of their well-being.
- **Early Disease Detection:** Wearables have the ability to spot minute changes in health measurements that might point to the beginning of a health problem. For instance, variations in sleep patterns may indicate sleep problems, whereas irregular heartbeat rhythms may indicate an arrhythmia. Early detection of these changes allows patients and healthcare professionals to take action before the situation gets worse.
- **Chronic Disease Management:** For people who are managing chronic diseases like diabetes, hypertension, or obesity, wearable sensors are very helpful. In order to enable more individualized treatment plans and prompt interventions, patients can track pertinent health measures and share this data with their healthcare professionals.
- **Medication Adherence:** In order to make sure that people take their prescribed drugs on time, several wearables can act as medication reminders. When it's time to take a tablet, these gadgets can warn patients and caregivers, enhancing adherence and treatment effectiveness.

- **Lifestyle Modification:** Real-time feedback on physical activity, calorie expenditure, and sleep patterns can encourage people to adopt healthier lives, which will improve their general health.
- **Telehealth Integration:** Platforms for telehealth and telemedicine can easily accommodate the health data collected by wearables. During virtual consultations, patients can share their data with healthcare professionals, giving them useful information for making better decisions.
- **Personalized Medicine:** Data from wearable sensors is useful in the developing field of personalized medicine. Healthcare professionals can customize treatment regimens to a patient's particular needs and genetic predispositions by combining genetic information, lifestyle data, and health indicators.
- **Research and Public Health:** Data from wearable sensors that have been aggregated and anonymized can be a goldmine for public health and medical research projects. This information can be used to spot patterns, disease outbreaks, and geographic regions that require specific actions.
- **Data Security and Privacy:** Strict data protection regulations are necessary to guarantee the reliability and security of wearable health data. Healthcare laws that guarantee data security and confidentiality, like HIPAA in the US, regulate how personal health information is handled.
- **Empowered Patients:** The empowerment of patients may be the most important effect of wearable sensors and health data integration. People now have more access to their health information, allowing them to actively manage their well-being and make educated health decisions.

Table 2.1 provides an overview of different types of wearable sensors and the health data they can collect, as well as some key considerations:

2.3.3 BIG DATA ANALYTICS FOR PERSONALIZED TREATMENT AND EARLY DISEASE DETECTION

The introduction of Big Data analytics has ushered in a disruptive age in the field of healthcare. Healthcare experts are now able to provide individualized treatment and early disease detection thanks to the power of large and diverse datasets, significantly altering the field of medicine [22] and patient care.

- **Personalized Treatment:** Highly individualized treatment plans can be created in the healthcare industry with the help of Big Data analytics. Healthcare professionals can custom-make treatments for each patient based on their genetic makeup, medical history, lifestyle, and real-time health data. The use of precision medicine guarantees that patients receive the best treatments possible with the fewest side effects.
- **Early Disease Detection:** Early disease discovery can, in some situations, save lives and dramatically improve treatment outcomes. Big Data analytics can spot minute trends and irregularities in patient data, allowing for the

TABLE 2.1

Types of Wearable Sensors Available

Wearable Sensor Type	Health Data Collected	Key Considerations
Heart rate monitor	Heart rate, heart rate variability	Accuracy, comfort, battery life
Accelerometer	Physical activity, motion, steps	Sampling rate, placement, battery life
GPS tracker	Location, distance, route	Battery life, GPS accuracy
ECG sensor	Electrocardiogram data	Accuracy, data storage, comfort
Blood pressure monitor	Blood pressure, heart rate	Calibration, cuff size, accuracy
Temperature sensor	Body temperature	Calibration, skin contact, power consumption
SpO$_2$ (oxygen saturation) sensor	Blood oxygen level	Accuracy, sensor placement
Sleep tracker	Sleep duration, quality	Comfort, battery life, data analysis
Glucose monitor	Blood glucose levels	Accuracy, calibration, data connectivity
Sweat sensor	Sweat composition	Calibration, skin contact, data analysis
UV radiation sensor	UV exposure	Accuracy, sun protection recommendations
Respiratory rate monitor	Breathing rate	Accuracy, sensor placement, battery life
Body composition analyzer	Body fat, muscle mass	Accuracy, user profile setup
Environmental sensor	Air quality, temperature	Sensor calibration, data interpretation
Blood alcohol sensor	Blood alcohol level	Accuracy, user education
EMG (electromyography) sensor	Muscle activity	Sensor placement, data processing
GSR (galvanic skin response) sensor	Skin conductance	Sensor calibration, data analysis

early diagnosis of ailments like cancer, diabetes, and cardiovascular issues. With this proactive strategy, healthcare moves away from a reactive one and toward one that emphasizes early detection and prevention.

- **Predictive Analytics:** Predictive analytics, a crucial part of Big Data in healthcare, forecasts health trends and disease risks using historical data and machine learning algorithms. By identifying high-risk individuals and putting preventive measures in place, healthcare practitioners can lessen the total burden of disease.
- **Real-time Monitoring:** Real-time health data, such as heart rate, blood pressure, and activity levels, are collected via wearable technology and IoT sensors. By processing this constant stream of data, Big Data analytics enables healthcare professionals to remotely monitor patients and quickly react to any deviations from typical health markers.
- **Drug Discovery and Development:** The analysis of enormous genomic and biochemical data sets using Big Data analytics speeds up the drug discovery process. This lessens the need for trial-and-error methods and

improves the outcomes of treatment by enabling the production of more focused and efficient drugs.

- **Healthcare Efficiency:** Big Data analytics optimizes resource allocation, lowers administrative costs, and boosts workflow effectiveness to expedite healthcare operations. This improves patient care while simultaneously bringing down the price of healthcare.
- **Population Health Management:** Public health organizations can employ Big Data analytics to monitor and control population health trends. To enhance general population health, this data-driven method assists in identifying health inequities, more efficiently allocating resources, and implementing focused interventions.
- **Data Security and Privacy:** Strong data protection procedures and adherence to healthcare standards, such as HIPAA in the United States, are essential to ensuring the security and privacy of patient data. These measures preserve patient confidentiality and confidence.
- **Patient Empowerment:** By giving patients access to their health data, Big Data analytics empowers patients. Patients are able to keep tabs on their health, make educated decisions regarding their care, and take an active role in their treatment regimens.
- **Research Advancements:** Big Data analytics is beneficial for the study of medicine. Large datasets are accessible to researchers, who can use them to understand illness processes, treatment effectiveness, and healthcare outcomes. This speeds up scientific advancement and raises the standard of care.

2.3.4 TRANSFORMING HEALTHCARE THROUGH IoT AND BIG DATA APPLICATIONS

The IoT and Big Data analytics are two technology behemoths that are coming together to change the healthcare sector [23]. By enabling more individualized therapies, early disease identification, and better patient outcomes, these advances are transforming the way healthcare is provided. Here are some examples of how IoT and Big Data are changing the healthcare landscape:

- **Real-Time Patient Monitoring:** IoT gadgets that monitor patients remotely and with wearable sensors continuously gather patient data. Real-time transmission of this information, which includes vital signs, activity levels, and medication adherence, is made to healthcare professionals. The early discovery of health problems is made possible by this ongoing monitoring, which also lowers readmissions to hospitals and enhances patient care.
- **Personalized Treatment:** Big Data analytics analyze enormous and varied datasets, including genetic data, EHRs, and patient histories. Healthcare professionals can develop treatment programs that are highly individualized by examining this data. In order to maximize therapeutic effectiveness and reduce side effects, this precision medicine approach makes sure that patients receive medicines that are specifically suited to their individual needs.

- **Early Disease Detection:** Inconspicuous trends and anomalies in patient data can be found using Big Data analytics, which may indicate the onset of diseases like cancer, diabetes, and heart disorders. The emphasis is shifted from managing diseases to avoiding them with this proactive approach to healthcare, potentially saving lives and lowering healthcare expenditures.
- **Predictive Analytics:** Big Data-driven predictive analytics foresees health trends and dangers using past health data and machine learning algorithms. These insights can be used by healthcare professionals to spot at-risk patients, act quickly, and put preventive measures in place. With this data-driven strategy, population health management is promoted, and the burden of chronic diseases is decreased.
- **Telemedicine and Remote Consultations:** Telemedicine solutions that enable remote consultations between patients and medical experts are made possible by IoT and Big Data. During virtual visits, patients can exchange their health information, allowing for better-informed decision-making and minimizing the need for in-person office visits.
- **Healthcare Efficiency:** Big Data analytics streamlines resource allocation, lowers administrative costs, and boosts workflow effectiveness to optimize healthcare operations. As a result, healthcare resources are better allocated and costs are reduced.
- **Drug Discovery:** The analysis of enormous genomic and biochemical data sets using Big Data analytics speeds up the drug discovery process. As a result, more precise and efficient drugs are created, increasing patient outcomes and lowering side effects.
- **Public Health Initiatives:** To manage population health trends, allocate resources wisely, and carry out focused interventions that will enhance general community health, public health organizations use Big Data analytics.
- **Patient Empowerment:** By gaining access to personal health information, patients are better equipped to actively manage their health, make educated decisions, and take part in their treatment regimens. This patient-centered methodology encourages participation and enhances medical results.
- **Data Security and Privacy:** Protecting patient data is of the utmost importance. Patient information is kept private and secure because of effective data security procedures and compliance with healthcare laws like HIPAA in the United States.

Some benefits of these applications of IoT and Big Data in healthcare are listed in Table 2.2.

2.4 SMART GRID

A modernized electrical distribution system known as a "smart grid" makes use of communication networks, intelligent devices, and advanced digital technologies to maximize the production, distribution, and consumption of electricity. Smart grids, as opposed to conventional grids, permit two-way communication between utilities

TABLE 2.2

Description and Benefits of Different Applications of IoT and Big Data in Healthcare

Applications	Descriptions	Benefits
Real-time patient monitoring	Use IoT devices to monitor patients' vital signs and health remotely	Early detection of health issues, reduced hospital readmissions, and improved patient outcomes
Personalized treatment	Big Data analytics analyze enormous and varied datasets, including genetic data, electronic health records (EHRs), and patient histories	Maximize therapeutic effectiveness, reduce side effects, and make sure that patients receive medicines that are specifically suited to their individual needs
Early disease detection	Use Big Data analytics to find anomalies in patients to indicate the onset of diseases like cancer, diabetes, and heart disorders	Potentially saving lives and lowering healthcare expenditures
Predictive analytics	Analyzing Big Data to predict disease outbreaks, patient admission rates, and treatment outcomes	Better resource allocation, and proactive healthcare planning
Telemedicine and remote consultation	Providing remote healthcare consultations and monitoring through IoT-enabled devices	Improved access to healthcare, reduced travel time and costs
Healthcare efficiency	Use Big Data analytics to allocate resources, lower administrative costs, and boost workflow effectiveness	Optimize healthcare operations, better allocation of healthcare resources, and ultimately reduce the costs
Drug discovery	Utilizing Big Data for drug discovery and clinical trials	Accelerated drug development, personalized medicine
Public health initiatives	Utilizing Big Data analytics to enhance the general community health	Manage population health trends, allocate resources wisely
Patient empowerment	By accessing personal health information, patients are better equipped to actively manage their health, make educated decisions, and take part in their treatment regimens	Encourages participation of patients and enhances medical results
Data security and privacy	Utilizing IoT and Big Data to keep patient information private and secure	Protection of patient data effectively due to data security procedures

and consumers, enabling real-time monitoring, control, and data sharing. Greater energy efficiency, fewer power outages, integration of renewable energy sources, and the capacity to support electric vehicles and other coming technologies are just a few advantages of this greater connectivity and automation. In order to fulfill the increasing needs of our modern society while reducing environmental impact and improving grid dependability, smart grids are essential to the transition to a more robust and sustainable energy infrastructure.

2.4.1   IoT Applications in Energy Management and Distribution

The IoT has changed the game for the energy industry by providing creative solutions to the problems associated with energy distribution and management. The way we produce, distribute, and use electricity is being revolutionized by IoT applications, which connect numerous devices, sensors, and systems across the power grid [24]. A closer look at how IoT is changing the energy landscape is provided below:

- **Smart Metering and Grid Monitoring:** Smart meters and sensors with IoT capabilities have been installed all throughout the grid, delivering real-time information on energy usage, voltage levels, and grid operation. Utility companies can monitor and operate the grid more effectively thanks to this data, which results in lower energy losses and more dependability.
- **Demand Response:** Demand-response solutions rely heavily on IoT, which enables utilities to interface with smart appliances and devices in homes and businesses. Utilities can remotely modify the operation of these devices to reduce energy consumption at times of high demand, ensuring grid stability and averting blackouts.
- **Distributed Energy Resources (DERs):** IoT makes it easier to integrate DERs into the grid, such as solar panels, wind turbines, and energy storage units. To maximize energy generation and storage, improve grid resilience, and encourage the use of renewable energy sources, these devices may be monitored and controlled remotely.
- **Predictive Maintenance:** IoT sensors on crucial grid elements, such as transformers and substations, gather information on the functionality of the equipment. This data is analyzed by machine learning algorithms to forecast maintenance requirements, which enables utilities to arrange repairs before breakdowns happen, minimizing downtime and cutting costs.
- **Grid Security:** Grid security is improved by IoT applications by continuously scanning for anomalies or online threats. Automated responses can be set off in the event of a breach to isolate the damaged locations and defend the grid against potential threats.
- **Energy Efficiency:** IoT devices in homes and buildings may monitor energy use patterns and provide customers with real-time feedback. Customers are better prepared to make energy-related decisions with this knowledge, which will ultimately lead to lower energy bills and a more sustainable system.
- **Electric Vehicle Integration:** Controlling and charging electric vehicles is made simpler by the IoT. Smart charging stations can connect to the grid in order to balance loads, lengthen charging times, and support the growing usage of EVs without unduly taxing the infrastructure.
- **Grid Flexibility:** Utility companies may now remotely operate grid assets thanks to IoT technologies, increasing system flexibility. This flexibility is crucial to accept intermittent renewable energy sources and manage system congestion.

- **Data Analytics:** Advanced data analytics are used to handle the enormous amounts of data provided by IoT devices. These data-driven insights help utilities improve grid performance, data-driven decision-making, and overall energy management.
- **Sustainability:** IoT applications support sustainability by enhancing energy efficiency, expanding the use of renewable energy sources, and lowering greenhouse gas emissions. They are essential to the shift to a more sustainable and clean energy ecosystem.

2.4.2 SMART METERS AND REAL-TIME ENERGY CONSUMPTION MONITORING

A whole new era of energy usage and control has begun with the introduction of smart meters. These cutting-edge gadgets, which have digital technology and two-way communication capabilities, have revolutionized how we monitor and manage our energy consumption. Smart meters' capacity to track energy use in real time has many advantages, from enabling consumers to make more informed decisions to improving grid sustainability and dependability [25], as shown in Figure 2.5.

- **Real-Time Data:** Consumers and utilities can both access real-time data on electricity consumption that smart meters continuously gather and transmit. This information offers a thorough understanding of energy trends since it includes specifics about when and how energy is used.
- **Empowered Consumers:** Customers can more carefully monitor their electricity usage if they have access to real-time energy data. They may now detect energy-hungry appliances, modify their usage patterns, and make wise decisions to cut down on energy wastage and utility costs, thanks to this newly discovered transparency.
- **Peak Load Management:** Utilities may manage peak electricity demand more effectively by utilizing real-time data from smart meters. In times of excessive demand, they can alert smart appliances or change electricity prices to encourage load-shifting and ease system stress.

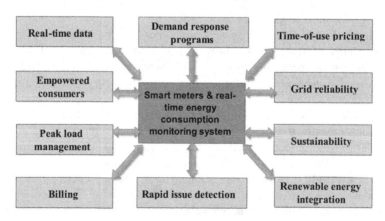

FIGURE 2.5 Smart meters and real-time energy consumption monitoring.

- **Billing Accuracy:** Manual meter readings are no longer necessary thanks to smart meters, which also provide precise and unchangeable invoicing. In addition to helping consumers, this simplifies utility administrative procedures.
- **Rapid Issue Detection:** Smart meters can detect anomalies in electricity consumption patterns, such as sudden spikes or drops, which may indicate equipment malfunctions or tampering. Utilities can respond promptly to such issues, ensuring the integrity of the grid.
- **Renewable Energy Integration:** Real-time energy consumption data aids in the integration of renewable energy sources like solar and wind. Consumers can align their energy usage with the availability of clean energy, maximizing their use of renewables and reducing reliance on fossil fuels.
- **Sustainability:** By empowering consumers to make more energy-efficient choices, smart meters contribute to sustainability efforts. Reduced energy consumption means lower greenhouse gas emissions and a smaller carbon footprint.
- **Grid Reliability:** Real-time data from smart meters helps utilities identify and respond to power outages more swiftly. This enhances grid reliability and minimizes downtime for consumers.
- **Time-of-Use Pricing:** Utilities can implement time-of-use pricing plans based on real-time data. Off-peak electricity pricing is available to consumers, further encouraging them to conserve energy.
- **Demand-Response Programs:** Utilities develop demand-response programs, which encourage customers to use less electricity during periods of high demand, using real-time data. The grid is stabilized as a result, and participants frequently benefit from decreased bills.

2.4.3 GRID STABILITY AND PREDICTIVE MAINTENANCE LEVERAGING BIG DATA ANALYTICS

Grid stability and preventive maintenance are crucial in the field of modern energy management to guarantee the dependable and effective distribution of electricity. Big Data analytics has become a revolutionary tool for reaching these goals [26,27]. Here's a closer look at how the power of Big Data analytics benefits grid stability and predictive maintenance:

2.4.3.1 Grid Stability

The ability of an electrical power system to maintain a consistent and balanced frequency and voltage under a variety of circumstances and loads is referred to as grid stability. Blackouts, brownouts, and disruptions in the energy supply can result from grid instability. By doing the following, Big Data analytics significantly contributes to grid stability:

- **Real-Time Monitoring:** Data collected by sensors and other devices placed across the grid is continuously processed using Big Data analytics. Real-time monitoring enables the quick identification of anomalies, fluctuations, or potential problems that can jeopardize the stability of the grid.

- **Predictive Analytics:** Predictive analytics forecasts probable grid instability triggers, such as peak demand periods, equipment failures, or weather-related difficulties, by examining past data and patterns. Utility companies can then take preventive action to avoid instability.
- **Load Forecasting:** Utilities can accurately forecast future electricity consumption with the use of Big Data analytics. With this knowledge, utilities can modify production and distribution to better meet customer demand, lowering the chance of overloads or underutilization.
- **Dynamic Grid Control:** By altering voltage levels, rerouting electricity, and controlling energy flows, advanced analytics can optimize grid operations in real time. These dynamic control techniques guarantee grid stability even in the face of unforeseen circumstances.

2.4.3.2   Predictive Maintenance

A data-driven strategy for maintaining grid equipment and infrastructure is called predictive maintenance. Utilities employ Big Data analytics to forecast when equipment is likely to break rather than depending on set timetables or responding to faults, allowing for prompt maintenance or replacement. This is how it goes:

- **Data Collection:** Data Gathering Transformers, substations, and electricity lines are just a few examples of the grid equipment that sensors and monitoring tools regularly gather data on. Variables including temperature, pressure, voltage, and current are included in this data.
- **Data Analysis:** This data is processed using Big Data analytics, which also finds patterns and anomalies that reveal information on the functionality and health of the equipment. Based on previous data and current conditions, machine learning algorithms can forecast possible failures.
- **Condition Monitoring:** Predictive maintenance programs continuously check the equipment's condition. Maintenance teams are notified to conduct inspections or repairs when specific thresholds or warning signs are found.
- **Reduced Downtime:** Utilities can dramatically cut downtime and service interruptions by recognizing and resolving problems before they result in equipment breakdowns. This lowers the cost of repair and replacement while also increasing grid reliability.
- **Asset Optimization:** Predictive maintenance guarantees optimal asset utilization of the grid's assets. Utilities enhance the return on investment in their infrastructure by extending the equipment's lifespan and minimizing unscheduled downtime.

2.4.4   DEMAND-RESPONSE MECHANISMS FOR EFFICIENT ENERGY UTILIZATION

Modern energy management must include demand-response technologies because they provide a proactive and dynamic method for maximizing energy use. These techniques enable customers and utilities to modify their energy consumption habits in response to changes in the supply and demand of electricity in real time [28]. By doing this, they support sustainability, grid stability, and more effective energy use.

- **Peak Load Management:** Effective management of peak electricity demand is one of the main advantages of demand-response mechanisms. In times of heavy demand, utilities might direct certain appliances or use pricing incentives to signal to customers to temporarily limit their electricity usage. As a result, the grid isn't put under as much stress, there are fewer chances of blackouts, and expensive infrastructure changes aren't required.
- **Grid Reliability:** Demand-response systems improve grid dependability by avoiding overloads and lowering the likelihood of grid outages during times of high demand or unanticipated events, such as extreme weather. Together, consumers and utilities can make sure that the grid runs efficiently, even in difficult situations.
- **Integration of Renewable Energy:** Grid operators may face difficulties due to the intermittent nature of renewable energy sources like solar and wind. Consumers can match their energy consumption to the availability of renewable energy thanks to demand-response devices. For instance, they can plan jobs that take a lot of energy during the times of day when solar panels are producing power, optimizing the usage of green energy.
- **Cost Savings:** Consumers can save money by taking part in demand-response initiatives. In order to encourage customers to switch their energy usage to periods when electricity is more inexpensive, utilities frequently provide lower electricity rates during off-peak hours. Consumer energy bills are decreased as a result, and utilities are also able to better allocate their energy resources.
- **Environmental Impact:** Demand-response mechanisms that effectively utilize energy result in lower greenhouse gas emissions. We can lessen our dependency on fossil fuels and lessen the negative effects of energy production on the environment by reducing peak demand and improving energy generation.
- **Grid Flexibility:** Grid operators have more flexibility in controlling energy supply and demand thanks to demand-response technologies. This adaptability is necessary to accommodate the increased adoption of electric vehicles, which can put additional strain on the grid, as well as intermittent renewable energy sources.
- **Consumer Empowerment:** Demand-response techniques provide consumers more control over their energy consumption, which increases their sense of empowerment. Consumers can make educated decisions about when and how to use electricity through real-time data and smart technologies, ultimately leading to more efficient and sustainable lifestyles.
- **Demand-Side Management:** By actively controlling energy use on the demand side, utilities can use demand-response techniques to balance the grid. Utility companies can optimize energy use to match supply, enhancing system stability by working with customers and businesses.

2.5 SMART INVENTORY SYSTEM

A technological advance in supply chain and inventory management is a smart inventory system. These systems offer real-time visibility into inventory quantities, locations, and conditions by utilizing cutting-edge technology like IoT sensors, RFID tags, and advanced data analytics [29]. Businesses may simplify processes, save carrying costs, eliminate stockouts, and improve reorder points with the help of smart inventory solutions. By automating inventory tracking and management, these solutions increase precision, decrease human error, and eventually boost overall supply chain efficiency, guaranteeing that companies can swiftly and accurately meet client requests.

2.5.1 IoT-Enabled Supply Chain Optimization

Supply chain management has entered a disruptive era thanks to the IoT, which provides unprecedented visibility, control, and efficiency across the entire logistical network [30]. IoT-enabled supply chain optimization makes use of a sizable ecosystem of networked gadgets, sensors, and data analytics to promote wiser decision-making, increase operational agility, and lower costs. Here are some specific examples of how IoT is transforming supply chain management:

- **Real-Time Visibility:** Real-time tracking of the location, state, and status of commodities and assets is made possible by IoT sensors and devices. Supply chain experts can precisely track shipments, inventory levels, and production processes thanks to this end-to-end visibility.
- **Predictive Analytics:** IoT data is fed into platforms for sophisticated analytics, providing insights into anticipated disruptions, delays, or supply chain bottlenecks. Businesses can foresee problems and take preventive measures using this proactive strategy to ensure smooth operations.
- **Inventory Optimization:** IoT sensors keep track of stock levels to prevent both overstocking and under stocking. This results in lower carrying costs, less waste, and better order fulfillment.
- **Condition Monitoring:** IoT sensors continuously monitor the status of delicate commodities like perishables or medicines. Alerts are sent out whenever the predetermined conditions are altered, allowing for quick action to stop spoilage or damage.
- **Supply Chain Traceability:** IoT improves transparency and traceability by producing an immutable record of a product's path from manufacturer to end-user. This is especially important for regulatory compliance and recalls in the food and pharmaceutical industries.
- **Fleet Management:** Real-time data on a vehicle's location, fuel consumption, and maintenance requirements is provided via IoT-connected vehicles. This information improves route planning, lowers fuel costs, and increases the useful life of fleet assets.
- **Demand Forecasting:** Demand forecasting is made easier and more reliable when IoT data is integrated with past sales and market data. As a result,

excess inventory and shortages are reduced as businesses are able to match production and inventory levels with actual market demand.

- **Sustainability:** By tracking energy use, emissions, and resource use, IoT helps sustainable supply chain operations. Businesses can pinpoint regions for development and lessen their influence on the environment.
- **Enhanced Customer Experience:** On-time deliveries are guaranteed by IoT-driven supply chain optimization, and order errors are decreased, which improves the entire experience for the consumer. Consumers who are pleased are more likely to become devoted, repeat consumers.
- **Cost Reduction:** Over time, IoT-enabled supply chain optimization results in significant cost savings by increasing operational effectiveness, decreasing waste, and avoiding disruptions.

2.5.2 REAL-TIME TRACKING AND MONITORING OF INVENTORY

Real-time inventory tracking and monitoring have developed into essential instruments in the fast-paced world of modern business and logistics in order to achieve effective supply chain management. The combination of cutting-edge technology like IoT, RFID, and sophisticated software solutions enables this capability [31]. Here's a closer look at how firms are managing their supply chains as a result of real-time inventory tracking and monitoring:

- **Unprecedented Visibility:** Businesses get an unmatched picture of their inventory at all times thanks to real-time tracking. Businesses may track the location, quantity, and status of their products in real time, whether they are in warehouses, distribution centers, in transit, using IoT sensors, RFID tags, or GPS technology.
- **Inventory Accuracy:** Automating tracking lowers the chance of human error and ensures inventory correctness. As a result, there are fewer instances of stockouts or overstocking, inventory levels are optimized, and order fulfillment is improved.
- **Efficient Replenishment:** Businesses can use real-time data on inventory levels to set up automatic reorder points or send out warnings when stock levels drop below predefined levels. By making the replenishment process more efficient and ensuring that products are only ordered when necessary, carrying costs are decreased.
- **Minimized Loss and Theft:** Continuous monitoring and real-time alerts enable swift action in case of inventory loss or theft. This deters theft and helps recover stolen goods or mitigate losses more effectively.
- **Condition Monitoring:** For businesses dealing with sensitive or perishable goods, real-time tracking includes monitoring environmental conditions. Sensors can detect factors like temperature, humidity, or light exposure, ensuring that goods are stored and transported under optimal conditions to maintain their quality.
- **Supply Chain Visibility:** Real-time tracking goes beyond the boundaries of the warehouse. It gives organizations visibility into the whole supply

chain, from suppliers to customers, allowing them to plan ahead for delays, make educated decisions, and optimize travel routes.

- **Regulatory Compliance:** Real-time tracking helps assure compliance with storage and transit restrictions in sectors with strong regulatory requirements, such as the food and pharmaceutical industries. Processes for auditing and reporting are made simpler.
- **Customer Satisfaction:** Businesses can quickly and accurately fill consumer orders thanks to accurate inventory tracking. Customers receive their items on time and in the desired quantity, which improves customer happiness and loyalty.
- **Data-Driven Decision-Making:** Real-time tracking generates a wealth of data that is essential for data analytics. Businesses can learn more about demand trends, inventory trends, and operational efficiency, allowing them to make data-driven decisions for ongoing improvement.
- **Cost Reduction:** Real-time tracking ultimately results in cost savings and increased overall supply chain efficiency by optimizing inventory levels, lowering manual labor, and decreasing stock outs.

2.5.3 DATA-DRIVEN DEMAND FORECASTING AND INVENTORY MANAGEMENT

Data-driven demand forecasting as well as inventory management have become important tools for companies looking to optimize their operations, reduce costs, and improve customer satisfaction in the dynamic world of supply chain management [32]. These data-driven techniques make use of cutting-edge analytics and machine learning algorithms to tackle the power of historical data, real-time information, and market flow to decide on inventory levels and demand in a way that is both accurate and rational. Figure 2.6 shows how data-driven inventory management and demand forecasting will affect supply chain efficiency in the near future:

- **Accurate Demand Forecasting:** Data-driven demand forecasting develops extremely precise forecasts of future demand by utilizing previous sales data, market trends, seasonality patterns, and even outside variables like weather or economic indicators. With such accuracy, firms may match production and inventory levels to the actual needs of the market, thereby eliminating excess inventory and lowering the risk of stockouts.
- **Real-Time Data Integration:** Contemporary supplier databases, IoT sensors, point-of-sale systems, and inventory management systems all integrate real-time data from a variety of sources. By ensuring that inventory levels are continuously changed to reflect shifting market conditions, this continuous data flow enhances responsiveness to client requests.
- **Reduced Carrying Costs:** Businesses can dramatically lower carrying costs related to surplus stock by optimizing inventory levels based on precise demand projections. As a result, money is freed up to be invested elsewhere in the company or utilized to increase the range of products available.
- **Enhanced Customer Satisfaction:** Successful business operations depend on rapidly and consistently meeting consumer requests. By ensuring that

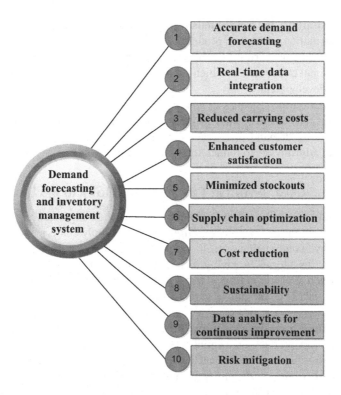

FIGURE 2.6   Data-driven inventory management and demand forecasting system affecting supply chain efficiency.

products are accessible when customers need them, data-driven inventory management increases customer happiness and loyalty.

- **Minimized Stockouts:** Data-driven inventory management aids organizations in preventing stockouts through real-time monitoring and predictive analytics. Product availability is always guaranteed by the capacity to foresee changes in demand and respond proactively, eliminating missed sales opportunities.
- **Supply Chain Optimization:** Data-driven strategies offer insights into the entire supply chain and go beyond the warehouse. Based on demand projections and real-time data, businesses can improve supplier relationships, transportation routes, and manufacturing schedules, increasing supply chain efficiency overall.
- **Cost Reduction:** Effective inventory management lowers the price of obsolescence, insurance, and storage. Additionally, it reduces the need for expensive last-minute production runs or accelerated shipment.
- **Sustainability:** By eliminating waste and the carbon footprint related to transit and storage, leaner inventory management helps with sustainability efforts. This fits with the increasing attention being paid to ecologically friendly corporate operations.

- **Data Analytics for Continuous Improvement:** Analytics can find a wealth of information in the data produced by data-driven inventory management. Businesses can learn more about consumer behavior, market trends, and operational effectiveness, which enables strategic decision-making and continual development.
- **Risk Mitigation:** Businesses can reduce risks related to supply chain interruptions, economic downturns, and unforeseen catastrophes by accurately anticipating demand and managing inventories. Planning ahead enables prompt adjustments to reduce future disruptions.

2.5.4 STREAMLINING INVENTORY PROCESSES USING IOT AND BIG DATA

The combination of IoT and Big Data analytics has ushered in a new era of accuracy and efficiency in inventory processes in the field of supply chain management [33]. This innovative combination of technology transforms how businesses manage and improve their inventories by giving them real-time insights, predictive capabilities, and data-driven decision-making. Here's a closer look at how IoT and Big Data are modernizing supply chain management and optimizing inventory processes:

- **Real-Time Inventory Visibility:** Real-time location, number, and condition of inventory products are continuously monitored by IoT sensors and devices. Businesses are now able to watch inventory movements across the whole supply chain, from manufacturing facilities to distribution hubs and retail outlets.
- **Predictive Demand Forecasting:** To produce extremely accurate demand projections, Big Data analytics examines past sales data, market trends, and other pertinent elements. These forecasts assist companies in anticipating changes in demand, ensuring that inventory levels are tailored to satisfy client needs while reducing excess stock.
- **Condition Monitoring:** IoT sensors with environmental and condition monitoring capabilities make ensuring that perishable or sensitive commodities are transported and stored in the best possible circumstances. Any departure from the predetermined parameters prompts quick notifications, enabling prompt responses to stop spoilage or damage.
- **Inventory Replenishment:** Business organizations can automate the replenishment process through real-time data integration. Systems can set up automated reorder points or produce alerts when inventory levels drop below specified thresholds to make sure that products are ordered precisely when needed and save down on carrying costs.
- **Supply Chain Traceability:** IoT-enabled inventory management procedures produce an unchangeable record of a product's path from manufacturer to end-user. Because it streamlines auditing and reporting procedures, improved traceability is especially important for sectors with strong regulatory requirements, like the pharmaceutical or food industries.
- **Reduced Human Error:** Automating inventory tracking and data entry reduces the possibility of human error. This results in increased order fulfillment, decreased stockouts, and improved inventory accuracy.

- **Data-Driven Decision-Making:** Data-driven decision-making is built on the data produced by IoT and Big Data analytics. Businesses receive insights into demand trends, inventory trends, and operational efficiency, enabling strategic planning and ongoing improvement.
- **Cost Reduction:** Simplified inventory procedures lead to cost savings by lowering carrying costs, reducing stockouts, and raising overall supply chain effectiveness. Businesses can invest in other important initiatives and deploy resources more effectively.
- **Sustainability:** By decreasing waste and lowering the environmental impact of transportation and storage, leaner inventory operations support sustainability initiatives. This fits with the increased emphasis on supply chain procedures that are ecologically friendly.
- **Risk Mitigation:** IoT and predictive analytics' real-time insights assist companies in reducing the risk of supply chain interruptions, market volatility, and unforeseen events. 2.6 Smart City

A smart city is a futurist urban environment that harnesses technology and data-driven solutions to improve the quality of life for its residents while optimizing resource efficiency and sustainability. In a smart city, a range of interconnected systems, such as transportation, healthcare, energy, and public services, communicate and collaborate all the way through the IoT, sensors, and advanced analytics [34]. This synergy permits for real-time monitoring, analysis, and management of urban operations, leading to improved infrastructure, decreased congestion, lessened energy consumption, improved public safety, and a more responsive and citizen-centric approach to administration. Smart cities intend to create vibrant, efficient, and sustainable urban spaces that address the evolving requirements of their residents while reducing environmental impact and fostering innovation.

2.6 IOT-DRIVEN INFRASTRUCTURE AND SERVICES IN URBAN SETTINGS

In the ferocious march toward urbanization, cities worldwide are holding with the challenges of population growth, resource constraints, and the requirement for sustainable, efficient, and livable environments. Internet of things (IoT) isa transformative force that is accommodating urban landscapes by infusing them with data-driven intelligence [35]. Figure 2.7 shows how IoT-driven infrastructure and services are remodeling urban settings, paving the way for smarter cities:

- **Smart Transportation:** IoT sensors installed in vehicles, roads, and traffic signals facilitate real-time traffic monitoring and congestion management and improve traffic flow. This not only decreases commute times but also lowers carbon emissions and improves road safety.
- **Efficient Energy Management:** Smart grids, powered by IoT, enable more efficient energy distribution and consumption. Buildings equipped with IoT sensors adjust lighting, heating, and cooling based on occupancy and environmental conditions, reducing energy waste.

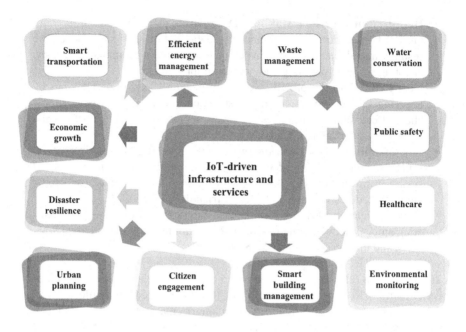

FIGURE 2.7   IoT-driven infrastructure and services used for urbanization.

- **Waste Management:** IoT-driven waste bins equipped with sensors signal when they are full, optimizing waste collection routes and reducing fuel consumption. This leads to cost savings and a cleaner urban environment.
- **Water Conservation:** IoT sensors monitor water quality, detect leaks, and manage irrigation systems more efficiently. These measures ensure a sustainable water supply and reduce water wastage in urban areas.
- **Public Safety:** Smart surveillance cameras, connected to IoT networks, enhance public safety by providing real-time monitoring and rapid response to incidents. This technology helps prevent crime and improve emergency response times.
- **Healthcare:** IoT-enabled healthcare devices and telemedicine services enhance access to healthcare in city areas. Remote monitoring and data sharing aid early disease detection and also efficient healthcare delivery.
- **Environmental Monitoring:** IoT sensors keep an eye on weather patterns, pollution levels, and air quality, enabling cities to develop data-driven policies for preserving the environment and urban planning.
- **Smart Building Management:** To improve building operations, IoT systems monitor and control the lighting, HVAC, and security systems. This reduces energy consumption while enhancing passenger comfort.
- **Citizen Engagement:** IoT-powered systems and apps enable citizens to take part in local government. They have the ability to report problems, use government services, and offer feedback, promoting a more adaptable and citizen-focused municipal administration.

- **Urban Planning:** IoT data is incredibly useful for urban planners. It contributes to more effective and sustainable urban growth by informing decisions regarding land use, traffic control, and infrastructure development.
- **Disaster Resilience:** IoT sensors are able to identify earthquake activity, flooding, and other natural calamities, enabling early alerts and quick action. This technique improves a city's ability to withstand unanticipated events.
- **Economic Growth:** Smart cities attract businesses and innovation hubs, fostering economic growth and job creation. The efficient infrastructure and improved quality of life in these cities make them appealing places to live and work.

2.6.1 Intelligent Waste Management and Environmental Monitoring

In our rapidly urbanizing world, managing waste and safeguarding the environment have become paramount concerns. Fortunately, the integration of intelligent technologies, such as the IoT and data analytics, is revolutionizing waste management and environmental monitoring practices in urban areas. This innovative approach not only enhances efficiency and sustainability but also contributes to cleaner, healthier, and more livable cities [36]. Here's a closer look at how intelligent waste management and environmental monitoring are leading the charge toward a greener urban future:

- **Smart Waste Bins:** IoT-equipped waste bins, featuring sensors that detect fill levels, streamline waste collection operations. These bins signal when they need emptying, optimizing collection routes, reducing fuel consumption, and minimizing the environmental impact of waste disposal.
- **Waste Sorting:** Intelligent waste sorting facilities use automation and AI to separate recyclables from non-recyclables more efficiently. This process reduces landfill waste, conserves resources, and promotes recycling.
- **Environmental Sensors:** IoT sensors monitor air quality, pollution levels, and weather conditions in real time. This data provides valuable insights for environmental management, helping cities address air and water pollution, and facilitating early responses to environmental hazards.
- **Water Quality Monitoring:** IoT-driven sensors continuously assess water quality in rivers, lakes, and reservoirs. This data aids in the early detection of pollutants, ensuring a safe and sustainable water supply for urban populations.
- **Energy-Efficient Waste Facilities:** IoT and data analytics optimize the energy consumption of waste treatment plants and incinerators. These facilities can adjust operations based on real-time data, reducing energy waste and operating costs.
- **Predictive Maintenance:** IoT sensors on waste management equipment and vehicles enable predictive maintenance. By detecting equipment issues early, cities can reduce downtime, extend the lifespan of machinery, and save on maintenance costs.

- **Environmental Impact Assessment:** Data analytics and modeling tools assess the environmental impact of urban development projects. This information guides urban planners in making sustainable decisions, minimizing ecological disruption.
- **Public Awareness:** Smart city apps and platforms engage citizens in waste reduction and environmental protection efforts. These platforms provide information on recycling, encourage responsible consumption, and empower individuals to report environmental issues.
- **Disaster Preparedness:** Environmental monitoring through IoT sensors can detect natural disasters like floods, wildfires, or landslides. This data enables early warnings and proactive measures, reducing the impact of disasters on urban areas.
- **Sustainable Urban Planning:** Intelligent waste management and environmental monitoring data inform urban planning decisions, helping cities create green spaces, optimize land use, and develop infrastructure that promotes sustainability.
- **Circular Economy:** By efficiently managing waste streams and promoting recycling, intelligent waste management contributes to the development of a circular economy, where resources are conserved, reused, and recycled, reducing the need for virgin materials.
- **Public Health:** Clean environments resulting from effective waste management and environmental monitoring contribute to improved public health by reducing exposure to pollution and hazardous waste.

2.6.2 Enhancing Public Safety Through IoT-Based Systems

In an increasingly interconnected world, public safety is a top priority for cities and communities. Leveraging the IoT, innovative IoT-based systems are transforming the way public safety is managed and delivered. These systems use sensors, data analytics, and real-time monitoring to create safer urban environments, respond swiftly to emergencies, and prevent incidents [37]. Figure 2.8 shows how IoT-based systems are enhancing public safety:

- **Real-Time Surveillance:** IoT-enabled cameras and sensors are deployed strategically throughout urban areas to provide continuous surveillance. These devices can detect suspicious activities, monitor traffic, and even identify potential hazards like fires or chemical leaks. Real-time video feeds and data analytics facilitate law enforcement to respond quickly to incidents and threats.
- **Predictive Policing:** Machine learning algorithms and IoT data can forecast crime hotspots and patterns based on past data. Law enforcement organizations are better able to manage their resources, discourage illegal behavior, and stop crime before it starts.
- **Emergency Response:** IoT-based technologies offer crucial information in times of need. For instance, sensors in buildings can find fires or structural problems, and intelligent traffic lights can prioritize emergency vehicle

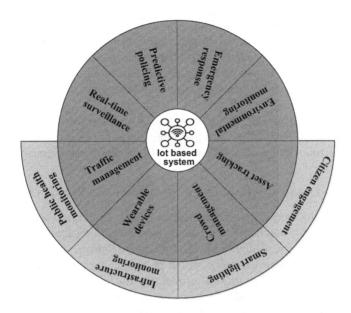

FIGURE 2.8   Enhancement of public safety in various domains by IoT-based system.

routes. These solutions guarantee more prompt and effective emergency responses, possibly saving lives.

- **Environmental Monitoring:** IoT sensors can identify environmental dangers like water contamination, air pollution, and severe weather. By enabling prompt evacuation or preventive action, early warnings and alarms contribute to the protection of public health and safety.
- **Asset Tracking:** IoT-based asset tracking solutions support the monitoring of important assets like equipment for emergency response or cars used for public transit. This guarantees their accessibility and can lower the possibility of theft or improper use.
- **Crowd Management:** IoT-based devices can track the density and movement of crowds during significant events or protests. Authorities can control crowds more safely and swiftly address any possible problems thanks to this knowledge.
- **Wearable Devices:** Wearable IoT devices that offer real-time health and location information can be helpful to first responders. These gadgets increase their security and make emergency situations easier to coordinate.
- **Traffic Management:** Traffic lights and sensors that are IoT-connected regulate traffic flow, ease congestion, and enhance road safety. Signals can be changed by adaptive traffic management systems in response to current circumstances, cutting down on the likelihood of accidents.
- **Public Health Monitoring:** Monitoring of public health: IoT can track the spread of illnesses, keep an eye on the air quality, and issue early warnings of potential health dangers. Public health organizations can better prevent illnesses and deal with emergencies by using this data.

- **Infrastructure Monitoring:** IoT sensors can keep an eye on the condition of vital infrastructure like tunnels and bridges. This proactive approach guarantees commuter safety and prevents accidents.
- **Smart Lighting:** By adjusting brightness based on current conditions, IoT-connected streetlights can increase visibility and safety in metropolitan areas while using less energy.
- **Citizen Engagement:** Through mobile apps or online platforms, IoT-based systems can enable citizens to report safety issues or crises, enabling shorter reaction times and encouraging a sense of community involvement.

2.6.3 DATA-DRIVEN GOVERNANCE FOR SUSTAINABLE AND RESILIENT CITIES

Cities all over the world are turning to data-driven governance as a potent tool to create sustainable and resilient urban environments in an age characterized by rising urbanization, the difficulties posed by climate change, and resource restrictions [38]. With this novel strategy, urban planning, resource management, and disaster preparedness are improved through the use of data analytics, real-time monitoring, and technology-driven decision-making. Here are some ways that data-driven governance is influencing the development of resilient and sustainable cities:

- **Informed Decision-Making:** Large datasets are gathered and analyzed as part of data-driven governance, which gives city officials invaluable knowledge on urban trends, population dynamics, and resource usage. Making decisions concerning resource allocation, infrastructure development, and transportation is now possible thanks to these insights.
- **Environmental Sustainability:** Cities can more efficiently monitor and manage their environmental impact thanks to data analytics. This entails monitoring waste production, water use, air quality, and emissions. Cities may implement sustainable policies and lessen their carbon footprint thanks to this data.
- **Resilience to Climate Change:** Cities may analyze their susceptibility to climate change and create risk-mitigation plans with the use of data-driven governance. Planning for early warnings and catastrophe response can benefit from real-time monitoring of weather trends and disaster preparedness data.
- **Efficient Resource Management:** Data-driven governance maximizes the use of resources. It assists communities in more effective management of their transportation, water, and energy systems, which lowers waste and operating expenses.
- **Smart Urban Planning:** Data analytics aid urban planners in creating cities that are more efficient and livable. The creation of transportation systems, green areas, and housing options is influenced by knowledge about people mobility, traffic patterns, and land use.
- **Public Services Improvement:** Data-driven governance improves the provision of public services. In order to ensure that resources are given where

they are most needed, city officials can use data to identify locations with a high demand for services like healthcare, education, and public transit.

- **Traffic Management:** Real-time data on traffic conditions and congestion empower cities to implement robust traffic management systems. This reduces commute times, eases congestion, and minimizes emissions from idling vehicles.
- **Citizen Engagement:** Data-driven governance promotes citizen participation in decision-making processes. Citizens can access and contribute data through digital platforms, providing valuable feedback and fostering a sense of community involvement.
- **Data Security and Privacy:** With the acceptance of data-driven governance, cities must prioritize data security and privacy. Robust cyber security scopes and policies are vital to protect sensitive information and maintain public trust.
- **Sustainable Infrastructure:** Data analytics direct the growth of sustainable infrastructure projects, like energy-efficient buildings, renewable energy sources, and green technologies, contributing to long-term sustainability.
- **Disaster Resilience:** Data-driven governance helps cities in disaster preparedness as well as response. Real-time info on weather, seismic activity, and emergency resources facilitate rapid, well-informed reactions to natural or man-made disasters.
- **Accountability and Transparency:** The usage of data endorses accountability and transparency in governance. Citizens can record the progress of projects, budgets, and policy implementations, holding officials responsible for their actions.

2.7 CHALLENGES AND FUTURE PROSPECTS

The unification of IoT and Big Data analytics has guided in a new era of innovation and possibilities, but it is not beyond its challenges. Managing the huge volume of data generated by IoT devices is a horrible task, demanding robust data storage, processing, and security solutions. Privacy considerations and data breaches are ever-present liabilities, demanding stringent measures to safeguard sensitive information. Additionally, interoperability topics between variety of IoT devices and platforms can prevent seamless data sharing and analysis. However, the expectations for the future are promising. Improvements in edge computing will enable real-time data processing at the source, decreasing latency and bandwidth constraints. AI and machine learning will further enhance data analytics capabilities, unlocking deeper insights and predictive capabilities. As IoT ecosystems continue to expand and mature, addressing these challenges will pave the way for more efficient, secure, and data-driven applications across industries, from healthcare and smart cities to supply chain management and beyond.

2.7.1 SECURITY AND PRIVACY CONCERNS IN IoT AND BIG DATA APPLICATIONS

As the IoT and Big Data analytics continue to revolutionize industries and daily life, they bring forth a host of security and privacy concerns that demand careful attention. The massive influx of data generated by IoT devices, coupled with the storage and analysis of this data, creates a fertile ground for potential vulnerabilities and threats. Here are some of the key security and privacy concerns in IoT and Big Data applications:

- **Data Privacy:** The sheer volume of data collected by IoT devices, ranging from personal health information to home automation preferences, raises significant concerns about individual privacy. Unauthorized access to this data can result in identity theft, surveillance, or misuse of personal information.
- **Data Breaches:** IoT devices are prone to hacking, and once agreed, they can serve as entry points for cyber-attacks. These gaps can lead to the disclose of sensitive data, financial losses, and even physical harm in some cases.
- **Lack of Standardization:** The IoT landscape is broken, with devices and platforms from various manufacturers repeatedly lacking standardized security protocols. This heterogeneity upsets efforts to set up consistent security measures across the IoT ecosystem.
- **Inadequate Authentication:** Weak or default authentication means in IoT devices can be used by malicious actors to gain unauthorized access. Strengthening authentication processes is vital to avoid unauthorized control or data theft.
- **Distributed Denial of Service (DDoS) Attacks:** IoT devices, when compromised, can be weaponized to commence massive DDoS attacks, upsetting critical online services and networks.
- **Supply Chain Vulnerabilities:** Security vulnerabilities may be initiated at any point in the supply chain, from tool manufacturing to software development. Ensuring the integrity of IoT components is vital.
- **Regulatory Compliance:** Stricter data protection regulations, such as the General Data Protection Regulation (GDPR), need organizations to realize robust security measures and transparent data handling practices, putting complexity to IoT as well as Big Data projects.
- **Insider Threats:** Insider threats, whether intentional or accidental, cause risks to IoT as well as Big Data security. Employees with access to sensitive data can compromise it, either through negligence or malicious intent.
- **Data Encryption:** Ensuring end-to-end encryption for data transmitted between IoT devices and data centers is vital to protect data from interception and tampering.
- **IoT Device Lifecycle Management:** Managing the security of IoT devices throughout their lifecycle, from deployment to retirement, is a complex task. Often, devices become outdated and unsupported, leaving them vulnerable to attacks.
- **Patch Management:** Keeping IoT devices up to date with security patches and updates can be challenging, as many devices lack automated update mechanisms. This creates a vulnerability window that malicious actors can exploit.

2.7.2 SCALABILITY AND INTEROPERABILITY CHALLENGES

While the IoT and Big Data offer immense potential for transforming industries and enhancing our daily lives, they also present significant scalability and interoperability challenges. These challenges can hinder the seamless expansion of IoT ecosystems and the integration of diverse data sources for effective Big Data analytics. Here, we delve into the key issues surrounding scalability and interoperability in IoT and Big Data applications:

2.7.2.1 Scalability Challenges

1. **Data Volume:** IoT devices generate an unprecedented volume of data. As the number of connected devices continues to grow exponentially, managing and processing this data efficiently becomes a formidable challenge.
2. **Data Storage:** Storing the massive amounts of data generated by IoT devices requires scalable and cost-effective solutions. Traditional databases may struggle to handle the sheer volume and variety of data types.
3. **Data Processing:** Analyzing vast datasets in real-time or near-real-time demands scalable computing resources. Ensuring that processing capabilities can keep pace with data influx is critical for timely insights.
4. **Network Infrastructure:** Scalable and reliable network infrastructure is essential for transmitting data between IoT devices and data centers. Network congestion or bottlenecks can impede data flow.
5. **Resource Constraints:** Many IoT devices have limited processing power, memory, and battery life. Scaling these devices while maintaining efficiency and performance is a significant challenge.

2.7.2.2 Interoperability Challenges

1. **Device Heterogeneity:** IoT ecosystems comprise devices from various manufacturers, often using different communication protocols and data formats. Ensuring these devices can communicate and cooperate seamlessly is a complex task.
2. **Data Standardization:** IoT devices often produce data in various formats. Standardizing data formats and communication protocols is crucial for enabling interoperability and data integration.
3. **Integration with Legacy Systems:** Many organizations have existing legacy systems that need to coexist with IoT solutions. Integrating these systems with modern IoT technologies can be challenging.
4. **Security and Identity Management:** Establishing trust and secure communication between diverse devices is essential. Managing identities, access control, and encryption keys across a heterogeneous IoT environment can be complex.
5. **Scalable APIs:** Developing scalable and reliable Application Programming Interfaces (APIs) that facilitate data exchange and communication between IoT devices and applications is vital for interoperability.
6. **Data Integration:** Combining data from various IoT sources with traditional data sources in Big Data analytics can be complicated. Data integration solutions must be robust and flexible.

2.7.3 Promising Future Directions for IoT-Driven Smart Environments

The realm of IoT-driven smart environments is continuously evolving, and as technology advances, promising future directions are emerging that hold the potential to reshape our connected world in profound ways. These innovations promise to enhance efficiency, sustainability, and the quality of life for individuals and communities. Here, we explore some of the exciting avenues that IoT-driven smart environments are heading toward:

- **5G Connectivity:** The rollout of 5G networks will revolutionize IoT connectivity by providing ultra-fast, low-latency connections. This will enable real-time data transmission and unlock new possibilities for applications like autonomous vehicles, telemedicine, and augmented reality in smart environments.
- **Edge Computing:** Edge computing, where data is processed closer to the source (IoT devices), will become more prevalent. This decreases latency, conserves bandwidth, and increases real-time decision-making, making IoT-driven applications faster and more responsive.
- **AI and Machine Learning Integration:** Smart environments will raising leverage AI and machine learning algorithms to analyze the large volumes of data provided by IoT devices. This will facilitate predictive analytics, anomaly detection, and personalized experiences in areas such as healthcare, transportation, and energy management.
- **Sustainable Practices:** IoT-driven smart environments will play a critical role in supporting sustainability. Energy-efficient buildings, optimized transportation systems, and intelligent resource management will help to minimize environmental impact as well as conserve resources.
- **Enhanced Healthcare:** IoT will carry on changing healthcare by enabling remote patient monitoring, personalized treatment plans, and early disease detection. Wearable devices and IoT-enabled medical equipment will grow to be more integrated with healthcare systems.
- **Smart Agriculture:** IoT-driven agriculture will optimize crop management, reduce resource use, and enhance yields. Sensors and drones will give real-time data on soil conditions, weather patterns, and crop health, set aside for precision agriculture.
- **Enhanced Security:** IoT-driven security systems will become more complicated, with AI-driven surveillance, biometrics, and threat detection. These devices will help to protect individuals, homes, and businesses more effectively.
- **Smart Transportation:** The future of transportation lies in autonomous vehicles, connected infrastructure, and shared mobility solutions. IoT will play a central role in making transportation safer, more capable, and environmentally friendly.
- **Citizen-Centric Governance:** IoT-driven data gathering and analysis will empower citizens to hold in urban governance. Smart cities will incorporate citizen feedback and inclinations into decision-making processes, developing a sense of community involvement.

- **Data Privacy and Security:** As IoT adoption develops, it will focus on data privacy and security. Future directions will engage the development of robust security measures, blockchain for data integrity, and advanced encryption techniques to protect sensitive information.
- **Interconnected Ecosystems:** IoT-driven smart environments will increasingly interconnect various ecosystems, like healthcare, transportation, and energy management, to give more holistic and efficient solutions.
- **Global Collaboration:** Collaboration among governments, industries, and academia will be necessary to address interoperability challenges, set up standards, and ensure the ethical and responsible deployment of IoT technologies in smart environments.

2.8   CONCLUSION

In the dynamic landscape of IoT and Big Data, the synergy of these technologies is propelling us toward a future filled with boundless possibilities. From smart cities and healthcare to energy management and supply chain optimization, the transformative impact of IoT and Big Data is undeniable. However, as we connect the power of these innovations, we must remain alert in addressing challenges linked to security, privacy, scalability, and interoperability. The promising future directions, marked by 5G connectivity, AI integration, sustainability, and citizen-centric governance, propose a glance at the potential yet to be unlocked. With responsible deployment, worldwide collaboration, and a commitment to ethical practices, we can guide this evolving landscape and make smarter, more efficient, and more inclusive environments that improve the lives of individuals and communities worldwide. The journey of utilizing the power of IoT and Big Data in smart environments has just started, and the possibilities for innovation and positive change are never-ending.

2.8.1   RECAPITULATION OF KEY FINDINGS

Throughout our survey of IoT and Big Data in smart environments, several essential findings have emerged. We have observed how IoT's proliferation of interconnected devices, coupled with Big Data analytics, is transforming industries and urban living. From healthcare advancements through remote monitoring to tenable urban planning driven by real-time data analytics, these technologies have enormous potential. However, they also offer challenges, including security and privacy concerns, scalability problems, and interoperability hurdles. Yet, the future holds promise, with the advent of 5G connectivity, AI integration, and a growing attention on sustainability and citizen engagement. As we give back on these findings, it's clear that responsible deployment, collaboration, and a commitment to ethical practices will be vital in harnessing the transformative power of IoT and Big Data for smarter, more capable, and more inclusive smart environments.

2.8.2   IMPLICATIONS OF IoT AND BIG DATA ADVANCEMENTS IN SMART ENVIRONMENTS

Advances in IoT and Big Data have significant and far-reaching effects on smart landscapes. These developments could greatly raise our standard of living, increase productivity, and promote sustainability in a variety of fields. Better health outcomes can be achieved in healthcare thanks to IoT-driven remote monitoring, which can change patient care and early disease identification. Through real-time data analytics, smart city efforts in metropolitan areas can lessen traffic, use less energy, and improve public safety. Through IoT-enabled tracking and inventory management, supply chains may become more effective and responsive while cutting costs and waste. These developments call for strong steps to secure sensitive data, but they also raise worries about data security and privacy. In the end, as we traverse this disruptive period, ethical issues, prudent deployment, and international cooperation will be paramount in harnessing the full potential of IoT and Big Data advancements in smart environments.

2.8.3   ENVISIONING A CONNECTED AND DATA-DRIVEN FUTURE

The possibilities are mind-blowing as we stand on the verge of a data-driven revolution in smart surroundings. Imagine a world where energy is used effectively, supply systems react in real time to demand variations, healthcare is tailored and preventive, and cities smoothly adjust to traffic patterns. In this future, data drives innovation, sustainability, and decision-making like the lifeblood of interconnected systems. However, while we envision this future, we must keep in mind the significance of ethical and responsible data use, preserving privacy and security, and promoting inclusivity in our networked society. Collaboration, creativity, and a dedication to using IoT and Big Data's revolutionary capacity for social good will distinguish the path to this future.

REFERENCES

[1] Y. Sasaki, "A survey on IoT Big data analytic systems: Current and future", *IEEE Internet of Things Journal*, vol. 9, no. 2, pp. 1024–1036, 2022.

[2] D. L. Andersen, C. S. A. Ashbrook, N. B. Karlborg, "Significance of big data analytics and the internet of things (IoT) aspects in industrial development, governance and sustainability", *International Journal of Intelligent Networks*, vol. 1, pp. 107–111, 2020.

[3] Y. Hajjaji, W. Boulila, I. R. Farah, I. Romdhani, A. Hussain, "Big data and IoT-based applications in smart environments: A systematic review", *Computer Science Review*, vol. 39, 2021, 100318. https://doi.org/10.1016/j.cosrev.2020.100318.

[4] G. Ikrissi, T. Mazri, "Iot-based smart environments: State of the art, security threats and solutions", *International Conference on Smart City Applications*, Turkey, pp. 279–286, 2021.

[5] D. Oladimeji, K. Gupta, N. A. Kose, K. Gindogan, L. Ge, F. Liang, "Smart transportation: An overview of technologies and applications", *Sensors*, vol. 23, no. 8, pp. 1–32, 2023. https://doi.org/10.3390/s23083880.

[6] S. C. Rai, S. Nayak, B. Acharya, V. C. Gerogiannis, A. Kanavos, T. Panagiotakopoulos, "ITSS: An intelligent traffic signaling system based on an IoT infrastructure", *Electronics*, vol. 12, no. 5, 1177 2023. https://doi.org/10.3390/electronics12051177.

[7] S. Damadam, M. Zourbakhsh, R. Javidan, A. Faoughi, "An intelligent IoT based traffic light management system: Deep reinforcement learning", *Smart Cities*, vol. 5, no. 4, 1293–1311 2022. https://doi.org/10.3390/smartcities5040066.

[8] V. Bali, S. Mathur, V. Sharma, D. Gaur, "Smart traffic management system using IoT enabled technology", *International Conference on Advances in Computing, Communication Control and Networking (ICACCCN), 2020.* https://doi.org/10.1109/ICACCCN51052.2020.9362753.

[9] R. Verma, P. Paygude, S. Chaudhary, S. Idate, "Real time traffic control using big data analytics", *2018 International Conference on Advances in Communication and Computing Technology (ICACCT)*, pp. 637–641, Sangamner, India 2018.

[10] F. Rau, I. Soto, D. Zabala-Blanco, C. Azurdia-Meza, M. Ijaz, S. Ekpo, S. Gutierrez, "A novel traffic prediction method using machine learning for energy efficiency in service provider networks", *Sensors*, vol. 23, no. 11, 4997 2023. https://doi.org/10.3390/s23114997.

[11] M. N. Tahir, P. Leviakangas, M. Katz, "Connected vehicles: V2V and V2I road weather and traffic communication using cellular technologies", *Sensors*, vol. 22, no. 3, pp. 1–14, 2022. https://doi.org/10.3390/s22031142.

[12] F. Jimenez, "Connected vehicles, V2V communications, and VANET", *Electronics*, vol. 4, no. 3, pp. 538–540, 2015.

[13] T. D. Toan, "Managing traffic congestion in a city: A study of Singapore's experiences", *International Conference on Sustainability in Civil Engineering*, Vietnam, 2019.

[14] C. Biyik, Z. Allam, G. Pieri, D. Moroni, M. Fraifer, E. O'Connell, S. Olariu, M. Khalid, "Smart parking systems: Reviewing the literature, architecture and ways forward", *Smart Cities*, vol. 4, pp. 623–642, 2021. https://doi.org/10.3390/smartcities4020032.

[15] E. Rowe, " The Los-Angeles automated traffic surveillance and control (ATSAC) system", *IEEE Transactions on Vehicular Technology*, vol. 40, no. 1, pp. 16–20, 1991.

[16] S. C. Zee, M. B. A. Dijkema, J. Laan, "The impact of inland ships and recreational boats on measured NOx and ultrafine particle concentrations along the waterways", *Atmospheric Environment*, vol. 55, pp. 368–376, 2012. https://doi.org/10.1016/j.atmosenv.2012.03.055.

[17] B. Joshi, K. Bhagat, H. Desai, M. Patel, J. K. Parmar, "A comparative study of toll collection systems in India", *International Journal of Engineering Research and Development*, vol. 13, no. 11, pp. 68–71, 2017.

[18] W. Zhao, X. Luo, T. Qiu, "Smart healthcare", *Applied Science*, vol. 7, no. 11, 1176 2017. https://doi.org/10.3390/app7111176.

[19] A. M. Ghosh, D. Halder, S. K. A. Hossain, "Remote health monitoring system through IoT", *5th International Conference on Informatics, Electronics and Vision (ICIEV)*, Bangladesh, 2016.

[20] S. Iranpak, A. Shahbahrami, H. Shakeri, "Remote patient monitoring and classifying using the internet of things platform combined with cloud computing", *Journal of Big Data*, vol. 8, pp. 1–22, 2021.

[21] P. P. Satve, S. Rani, N. Chinthamu, J. S. Mehta, K. G. S. Venkatesan, P. Mishra, "Wearable sensors and IoT integration towards personalized and pervasive healthcare", *European Chemical Bulletin*, vol. 12, pp. 4712–4723, 2023.

[22] D. Cirillo, A. Valencia, "Big data analytics for personalized medicine", *Current Opinion in Biotechnology*, vol. 58, pp. 161–167, 2019.

[23] A. Rejeb, K. Rejeb, H. Treiblmaier, A. Appolloni, S. Alghamdi, Y. Alhasawi, M. Iranmanesh, "The Internet of Things (IoT) in healthcare: Taking stock and moving forward", *Internet of Things*, vol. 22, 1–23 2023. https://doi.org/10.1016/j.iot.2023.100721.

[24] K. Mishra, S. Goyal, V. A. Tikkiwal, A. Kumar, "An IoT based smart energy management system", *4th International Conference on Computing Communication and Automation (ICCCA)*, pp. 1–3, Greater Noida, India 2018.

[25] C. Landi, G. D. Prete, D. Gallo, "Real-time smart meters network for energy management", *ACTA IMEKO*, vol. 2, no. 1, pp. 998–1003, 2012. https://doi.org/10.21014/acta_imeko.v2i1.51.

[26] Y. Zhang, T. Huang, E. F. Bompard, "Big data analytics in smart grids: A review", *Energy Informatics*, vol. 1, no. 8, pp. 1–24, 2018.

[27] N. Mostafa, H. Ramadan, O. Elfarouk, "Renewable energy management in smart grids by using big data analytics and machine learning", *Machine Learning with Applications*, vol. 15, 1–12 2022.

[28] K. Wohlfarth, E. Worrell, W. Eichhammer, "Energy efficiency and demand response - Two sides of the same coin?", *Energy Policy*, vol. 137, pp. 1–11, 2020.

[29] S. Paul, A. Chatterjee, D. Guha, "Study of smart inventory management system based on the internet of things (IOT)", *International Journal on Recent Trends in Business and Tourism*, vol. 3, no. 3, pp. 27–34, 2019.

[30] A. Rejeb, S. Simske, K. Rejeb, H. Treiblmaier, S. Zailani, "Internet of Things research in supply chain management and logistics: A bibliometric analysis", *Internet of Things*, vol. 12, pp. 1–16, 2020.

[31] S. B. Erlangga, A. Yunita, S. R. Satriana, " Development of automatic real time inventory monitoring system using RFID technology in warehouse", *International Journal on Informatics Visualization*, vol. 6, no. 3, pp. 636–642, 2022. https://doi.org/10.30630/joiv.6.3.1231.

[32] M. Seyedan, F. Mafakheri, "Predictive big data analytics for supply chain demand forecasting: Methods, applications, and research opportunities", *Journal of Big Data*, vol. 7, no. 53, pp. 1–22, 2020.

[33] L. He, M. Xue, B. Gu, "Internet-of-things enabled supply chain planning and coordination with big data services: Certain theoretic implications", *Journal of Management Science and Engineering*, vol. 5, no. 1, pp. 1–22, 2020.

[34] M. Talebkhah, A. Sali, M. Marjani, M. Gordan, S. J. Hashim, F. Z. Rokhani, "IoT and big data applications in smart cities: Recent advances, challenges, and critical ssues", *IEEE Access*, vol. 9, pp. 55465–55484, 2021.

[35] A. S. Syed, D. Sierra-Sosa, A. Kumar, A. Elmaghraby, "IoT in smart cities: A survey of technologies, practices and challenges", *Smart Cities*, vol. 4, pp. 429–475, 2021.

[36] S. T. Ikram, V. Mohanraj, S. Ramachandran, A. Balakrishnan, "An intelligent waste management application using iot and a genetic algorithm-fuzzy inference system", *Applied Science*, vol. 13, no. 6, pp. 1–29, 2023.

[37] M. Thibaud, H. Chi, W. Zhou, S. Piramuthu, "Internet of Things (IoT) in high-risk environment, health and safety (EHS) industries: A comprehensive review", *Decision Support Systems*, vol. 108, pp. 79–95, 2018.

[38] M. Rathore, A. Ahmad, A. Paul, S. Rho, "Urban planning and building smart cities based on the Internet of Things using big data analytics", *Computer Networks*, vol. 101, pp. 63–80, 2016.

3 Internet of Things Architecture's Importance in Healthcare Management

*Inderpal Singh, Mamta Devi,
and Arun Kumar Rana*

3.1 INTRODUCTION

The primary tenet of the Internet of Things (IoT) involves establishing seamless connectivity between intelligent entities, sometimes referred to as "things," and the Internet. This facilitates the transfer of data among many entities, hence enhancing the delivery of information to users in a more robust and secure manner. This development is expected to involve the widespread distribution of physical objects, such as computers and sensor actuators, equipped with distinct addresses and the capability to securely transmit data. These devices will facilitate the transfer of information ranging from routine daily activities to sensitive medical records [1–3].

The technology referred it and empowered seamless integration for numerous hardware with advanced technologies, enabling effortless connectivity, communication, sensing, interaction with the physical environment, and inter-device communication [19]. IoT stands as a visionary framework, uniting a network of interconnected entities encompassing individuals, objects, services, and networks, all operating harmoniously across diverse locations and timeframes [16].

In the ever-evolving landscape of IoT applications, healthcare emerges as one of the most compelling domains. This is attributed to its myriad medical applications, including but not limited to monitoring, personalized and exercise programs, efficient management of chronic diseases, and enhanced care for the elderly [16]. Moreover, IoT continues to unlock innovative possibilities in healthcare, transforming the way we approach and deliver medical services [4,5].

3.2 BACKGROUND AND SIGNIFICANCE OF IOT IN HEALTHCARE

The healthcare industry is undergoing a profound transformation through the versatile applications of IoT technologies and devices. This transformation offers numerous benefits to both patients and healthcare providers as the IoT continues to expand its footprint within healthcare.

DOI: 10.1201/9781032673479-4

Healthcare IoT encompasses a wide range of applications, including mobile medical apps and wearable devices that empower users to collect and monitor their health data. Moreover, this technology to effectively monitors the location and usage of medical equipment, staff, and patients in their care. Here, we present a compilation of various technologies poised for implementation within healthcare systems driven by IoT.

While the integration of IoT devices is revolutionizing healthcare, it also raises substantial security concerns for clinical engineering and corporate technology teams. These intelligent devices have the capacity to establish connections with global information networks, granting users access to information at any time and place.

3.3   ARCHITECTURE FOR IOT-ENABLED BIG DATA ANALYTICS

IoT architecture encompasses various systems derived from abstracting and identifying different IoT domains [6].

Within the ambit of this pioneering framework, we lay the cornerstone of a reference model that bridges the diverse verticals within the IoT domain. These verticals encompass the intricate domains of intelligent traffic management, the seamless operation of smart homes, the optimization of transportation systems, and the transformation of healthcare into a realm of intelligence.

This meticulously structured architectural framework, meticulously tailored to meet the rigorous demands of big data analytics, serves as not just a guiding light but a sophisticated roadmap for data abstraction [7,8]. It stands as a testament to the harmonious convergence of advanced technology and pragmatic utility, ushering in a unique era where interconnected possibilities flourish, and innovation thrives. However, it's worth noting that the prevailing architectural framework primarily centers around IoT in the context of communication. To the best of our knowledge, no prior research has ventured into the exploration of our innovative architecture, which ingeniously integrates IoT with the vast landscape of big data analytics. As depicted in Figure 3.3, our IoT architecture showcases sensor devices and components in the sensor layer, interlinked via wireless networks employing a diverse array of technologies, including radio-frequency identification (RFID), WiFi, ultrawideband, ZigBee, and Bluetooth. Enabling the communication bridge between the Internet and interconnected networks is an IoT gateway.

Zooming in at a higher echelon, our focus sharpens on big data analytics, where a substantial volume of sensor data finds its sanctuary in cloud-based systems, accessible through specialized big data analytics applications. These programs encompass not only API administration functionality but also a user-friendly graphical interface (dashboard), providing a seamless interaction with the processing engine.

This study introduces a pioneering approach that leverages a meta-model for the integration of elements within an IoT architecture. This concept seamlessly melds into a holistic digital enterprise architecture environment, facilitated by a semi-automated process. The overarching goal revolves around furnishing effective decision support for intricate business and architectural management tasks, including the development of evaluation systems and IT environments. Consequently, architectural decisions linked to IoT intricately intertwine with code applications, empowering stakeholders to grasp the symbiosis between business architecture management and IoT [9,10].

3.4 IOT ARCHITECTURE IN HEALTHCARE

In the realm of healthcare delivery within IoT, there exists a foundational structure consisting of three key layers, as detailed in reference [12]. Our intention here is not to offer a comprehensive breakdown of these layers; instead, we will offer a brief overview along with the health-related implications in the following sections.

In recent times, there has been a notable surge in the development of healthcare applications [11,15] within the realm of the IoT. The system has been proposed with the goal of addressing the needs of patients, healthcare professionals, and administrative staff. IoT applications are instrumental in augmenting accessibility by enabling real-time patient condition monitoring and facilitating effective medical emergency management, among other advantages. These applications empower patients to document and track their current health status, encompassing parameters such as blood pressure, glucose levels, cardiac performance, physical fitness, and other pertinent metrics. Subsequently, the collected data is transmitted to the relevant healthcare facility or physician. Healthcare institutions, including hospitals, are increasingly adopting IoT applications to deliver real-time services, allowing for precise patient location identification and continuous health problem monitoring (see Figure 3.1).

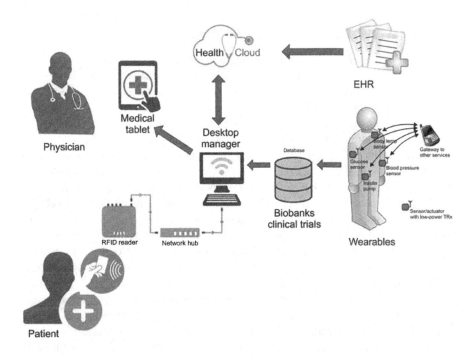

FIGURE 3.1   The IoT network.

3.5 PERCEPTION LAYER: SYSTEMS RESPONSIBLE FOR SENSING AND COLLECTING DATA

Perception and identification technologies stand as the foundational pillars of the IoT, representing a remarkable nexus of innovation. Sensors, these electronic marvels, possess the remarkable capability to discern and record changes within their immediate surroundings. This sensor category boasts an impressive diversity, featuring a myriad of sensor types, including but not limited to RFID sensors, infrared sensors, cameras, global positioning system (GPS) sensors, medical sensors, and sensors seamlessly integrated into smart devices.

Harnessing these sensor technologies enables the comprehensive perception of the world, encompassing object recognition, positional awareness, and geographic identification. What further elevates their significance is their innate ability to translate this wealth of information into digital signals, thereby optimizing the efficiency of network transmission, as elegantly discussed in references [12,13].

The transformative impact of sensor technologies extends into the realm of healthcare, where they enable real-time monitoring of therapies and streamline the collection of a diverse array of physiological indicators pertinent to patient care. This streamlined process expedites the diagnostic journey and facilitates the administration of top-tier medical treatment. It is worth noting the existence of numerous IoT sensor devices with the potential for life-saving interventions. However, a note of caution is in order, as not all of these devices have undergone rigorous clinical testing or demonstrated their safety and efficacy, underscoring the need for further research and scrutiny.

3.6 NETWORK LAYER: DATA COMMUNICATION AND STORAGE

The network layer of Internet of Things (IoT) technologies is a multifaceted realm, encompassing both wired and wireless networks. Within this intricate landscape, networks serve as the conduits for communication and storage of processed information, offering the flexibility of local or centralized data repositories. IoT, as a paradigm, places paramount importance on communication across a spectrum of frequencies, accommodating high, low, and medium ranges.

Within this spectrum, IoT technologies encompass various forms of short-range communication, including RFID, wireless sensor networks, Bluetooth, Zigbee, low-power Wi-Fi, and global system for mobile communications, as extensively detailed in reference [12]. Embracing high-frequency fourth-generation (4G) cellular networks amplifies communication capabilities, and the ongoing evolution toward fifth-generation (5G) networks further expands horizons, promising accessibility improvements. These advanced networks are poised to play a pivotal role in the burgeoning realm of IoT applications in healthcare, offering the tantalizing prospect of establishing dependable connections among numerous devices simultaneously, potentially scaling to thousands of devices concurrently [14].

The data conveyed through IoT networks can find its repository in two distinct locales: local, often decentralized storage, or transmission to a centralized cloud server. The utilization of cloud-based computing in healthcare bears a multitude of advantages, characterized by its ubiquitous availability, adaptability, and capacity for scalable data acquisition, storage, and transmission among cloud-connected devices [18]. This approach holds the promise of underpinning data-intensive electronic medical records (EMRs), patient portals, medical IoT devices (including smartphone applications), and the analysis of vast datasets that drive decision support systems and therapeutic strategies [5]. However, as the healthcare industry increasingly integrates cloud applications, the imperatives of substantiating their efficacy and safety, addressing security concerns pertaining to health data, ensuring the reliability and transparency of data access by third parties, and navigating potential challenges associated with data accumulation and latency issues due to geographical separation between IoT devices and data centers loom large.

The advent of decentralized data processing and networking strategies stands as a transformative proposition, poised to elevate the scalability of IoT applications in healthcare. The concept of the edge cloud, a contemporary cloud computing paradigm, empowers IoT sensors and network gateways to autonomously process and analyze data at the periphery, decentralizing computational tasks. This innovative approach effectively reduces the volume of data requiring transmission and centralized administration, a phenomenon expounded upon.

In a similar vein, the utilization of blockchain storage introduces a decentralized methodology for data storage, with autonomous blocks housing distinct sets of information. These blocks interlink in a symbiotic manner, forming a cohesive block structure. Consequently, a participant-governed network emerges, rendering third-party entities redundant [14]. Notably, certain platforms have embarked on the engineering of blockchain technology for medical applications. Nevertheless, the current landscape of research concerning the integration of edge cloud and blockchains within the healthcare sector remains relatively uncharted, underscoring the urgent need for comprehensive exploration in this promising domain.

3.7 APPLICATION LAYER

The application layer assumes a pivotal role, tasked with the nuanced interpretation and precise implementation of data, in addition to delivering application-specific services to end users, as elucidated in reference [12]. Within the intricate framework of the IoT, artificial intelligence (AI) emerges as a highly promising avenue, particularly within the realm of medical applications. The utilization of AI within scientific domains has witnessed a substantial surge, spanning diverse fields such as image analysis, natural language processing for text recognition, the design of therapeutic interventions, and the prediction of gene mutation expressions, as documented extensively [23].

AI wields the formidable capacity to scrutinize a trove of accessible EMR data, encompassing an individual's medical history, physical examination findings, laboratory results, medical imaging, and prescribed medications. Through the contextualization of this wealth of information, AI stands poised to generate potential treatment options and diagnostic choices. A vivid illustration of AI's prowess in healthcare is exemplified by IBM Watson. This AI system boasts the ability to comprehend both structured and unstructured textual data within EMRs. Moreover, it possesses the acumen to analyze medical images to discern significant and secondary observations.

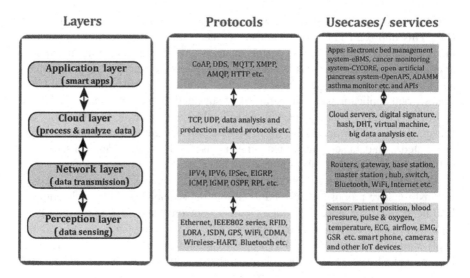

FIGURE 3.2   Layer-wise protocols and technologies of the IoT healthcare architecture.

Furthermore, IBM Watson adeptly curates relevant medical literature to furnish comprehensive responses to clinical inquiries, as articulated in reference [24].

The marriage of IoT technology with the application of deep machine learning augments the healthcare landscape, empowering practitioners to perceive hitherto imperceptible phenomena and ushering in novel and enhanced diagnostic capabilities. While the attainment of a perfect diagnostic confidence score of 100% remains an elusive goal, the amalgamation of machine capabilities with the seasoned experience of clinicians consistently elevates the overall performance of the healthcare system (Figure 3.2).

3.7.1 Healthcare Applications

Electrocardiogram (ECG) monitoring is essential as it can detect circulatory system disorders responsible for around 30% of mortalities. ECG records heart electrical activity, aiding in diagnosing arrhythmias, ischemia, and more. IoT-based solutions transmit this data to healthcare professionals. Blood pressure monitoring is crucial in preventing circulatory disorders, with IoT apps enabling remote communication between health posts and centers. Monitoring body temperature is vital for maintaining health equilibrium. IoT-integrated sensors in devices like TelosB can track temperature fluctuations and transmit data through the IoT. Oxygen saturation levels can be monitored non-invasively using pulse oximetry, which IoT applications can facilitate. IoT-based smart rehabilitation systems, discussed in reference [17], address the challenges of an aging population, including cerebrovascular accidents, by offering automated rehabilitation programs. Medication management is a significant challenge in public health, with IoT offering innovative solutions. Advancements in IoT have also led to the development of fully automated smart wheelchairs, improving mobility for individuals with disabilities (Figure 3.3).

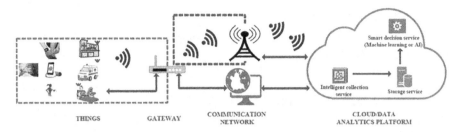

THINGS GATEWAY COMMUNICATION NETWORK CLOUD/DATA ANALYTICS PLATFORM

FIGURE 3.3 The framework of healthcare service.

One example of a framework that may be implemented in a practical application is the Remote Monitoring and Management Platform of Healthcare Information (RMMP-HI) [22]. This platform offers the capability to monitor and manage lifestyle diseases with the aim of achieving preventive and early detection objectives. Body medical sensors have the capability to record, remove, and modify data inside an IoT network. These sensors gather medical information pertaining to the human body and transmit it to a data-sharing center. Subsequently, the data is disseminated to medical personnel or hospital facilities in accordance with predefined regulations, including the transmission of urgent notifications to hospitals.

3.7.2 Security and Privacy Requirement of Medical Internet of Things (MIoT)

Ensuring the security and privacy of MIoT is paramount in modern healthcare. Healthcare organizations often lack adequate resources for this, even though MIoT systems are complex, comprising various components like medical information systems, gateways, and smart devices, all producing sensitive real-time data. To address this, several requirements must be met in MIoT system design. Confidentiality restricts access to authorized personnel, data integrity safeguards against unauthorized tampering, and data availability ensures timely access for authorized users. Resilience against attacks requires avoiding single points of failure and comprehensive protection frameworks. Data usability limits access to authorized individuals, access control mechanisms ensure security, and data auditing tracks record access for monitoring. Data authentication verifies data source and integrity. Sensitive patient data, including details on diseases and personal information, necessitates strong measures to prevent unauthorized access, even during data collection and interception. These considerations are vital for safeguarding MIoT in contemporary healthcare [19,20].

3.7.3 Security Challenges in Internet of Health Things (IoHT)

In order to address the emerging challenges posed by the IoT, it is imperative to use fresh counter practices that can effectively tackle the new concerns arising from this technology. This is necessary since traditional protection techniques are insufficient in ensuring the security and privacy of IoT devices and networks [21]. Several issues exist in protecting IoHT infrastructure. These challenges include:

3.7.4   ENERGY AVAILABILITY

A typical IoT network in the healthcare sector comprises compact medical devices that are equipped with limited battery capacity. These devices conserve energy by activating the power-saving mode when there is no need for sensor observations to be recorded. Furthermore, the central processing units (CPUs) exhibit reduced performance when there are no critical tasks to be done [21]. Hence, the presence of power constraints in IoT medical devices poses a significant issue in the search for an energy-conscious security solution.

3.7.5   DYNAMIC SECURITY UPDATES

There exists a significant imperative to ensure the regular updating of security protocols in order to mitigate the presence of high-risk vulnerabilities. Therefore, it is imperative to ensure that IoT medical devices are equipped with the latest security fixes. However, devising a methodology for expediting the installation of protective systems presents a challenging task.

3.7.6   MULTIPLICITY OF GADGETS

The healthcare devices encompassed under an IoHT system exhibit a wide range of capabilities, spanning from sophisticated personal computers to more basic RFID tags. These devices exhibit variations in terms of their computational capacity, power consumption, embedded software functionality, and storage capacity

3.7.7   SCALABILITY

The proliferation of IoT devices has exhibited a consistent upward trend, resulting in a significant increase in the number of devices connected to the global data network. Therefore, the task of developing a highly adaptable security strategy that effectively addresses security requirements while maintaining flexibility presents a significant difficulty.

3.7.8   MEMORY LIMITATIONS

The majority of IoHT system gadgets possess limited on-gadget memory, necessitating their use of an integrated operating system (OS), binary application, and system software.

3.7.9   COMPUTATIONAL LIMITATIONS

The efficiency of the CPU in IoT medical equipment is compromised due to the integration of low-speed processors [25]. Furthermore, primarily function as actuator/sensor systems [26]. Therefore, the task of finding a protective solution that minimizes resource use and consequently enhances safety is a challenging endeavor.

3.8　CONCLUSION

Prior to the advent of IoT technology, the means of communication between patients and medical personnel were tele-interactions and test-based conversations. There existed no alternative method through which medical professionals and healthcare facilities could consistently monitor the health status of patients and provide corresponding suggestions. The integration of IoT devices has facilitated the implementation of remote monitoring systems in the healthcare industry, hence enabling the maintenance of patients' well-being and safety. Additionally, these IoT-enabled gadgets have been shown to be crucial in enhancing the quality of care delivered by medical personnel. Furthermore, there has been an enhancement in patient happiness and engagement due to the simplification and increased efficiency of doctor-patient communication. In addition, remote monitoring of patients' health contributes to reducing the length of hospitalization and preventing readmissions. IoT significantly impacts the reduction of healthcare expenditures and the improvement of treatment outcomes. There is no doubt that the IoT is significantly transforming healthcare zone by expanding the realm of devices and enhancing communication between individuals, hence offering innovative healthcare solutions. The IoT spans a diverse array of applications and services in the healthcare sector, providing numerous advantages to medical professionals, patients, caregivers, families, healthcare institutions, and even insurance firms.

REFERENCES

1. Rehman, A.U., Khan, B.: Challenges to online education in Pakistan during COVID-19 & the way forward. *Soc. Sci. Learn. Educ. J.* 6, 503–512 (2021).
2. Thuburn, D.: WHO Has Finally Named the New Coronavirus, 12 February 2020. Available online: https://www.sciencealert.com/who-hasfinally-named-the-deadly-coronavirus (accessed on 20 May 2022).
3. Malik, S., Ahmad, S., Ullah, I., Park, D.H., Kim, D.: An adaptive emergency first intelligent scheduling algorithm for efficient task management and scheduling in hybrid of hard real-time and soft real-time embedded IoT systems. *Sustainability* 11, 2192 (2019).
4. Ahmad, S., Kim, D.: A multi-device multi-tasks management and orchestration architecture for the design of enterprise IoT applications. *Future Gener. Comput. Syst.* 106, 482–500 (2020).
5. Mehmood, F., Ullah, I., Ahmad, S., Kim, D.: Object detection mechanism based on deep learning algorithm using embedded IoT devices for smart home appliances control in CoT. *J. Ambient. Intell. Humaniz. Comput.* 19, 1–17 (2019).
6. Laplante, P.A., Laplante, N.: The internet of things in healthcare: Potential applications and challenges. *IT Prof.* 18, 2–4 (2016).
7. Pace, P., Aloi, G., Gravina, R., Fortino, G., Larini, G., Gulino, M.: Towards interoperability of IoT-based health care platforms: The INTER-health use case. In: *Proceedings of the 11th EAI International Conference on Body Area Networks*, Turin, Italy, 15–16 December 2016.
8. Liotta, A.: The cognitive NET is coming. *IEEE Spectr.* 50, 26–31 (2013).
9. Javed, F., Samiullah, K., Khan, A., Javed, A., Tariq, R., Matiullah, Khan, F.: On precise path planning algorithm in wireless sensor network. *Int. J. Distrib. Sens. Netw.* 14, 1–12 (2018).

10. Saeed, N., Bader, A., Al-Naffouri, T.Y., Alouini, M.S.: When wireless communication faces COVID-19: Combating the pandemic and saving the economy. *Front. Commun. Netw.* 1, 1–15 (2020).
11. Rodrigues, J.J., Segundo, D.B.D.R., Junqueira, H.A., Sabino, M.H., Prince, R.M., Al-Muhtadi, J., De Albuquerque, V.H.C.: Enabling technologies for the internet of health things. *IEEE Access* 6, 13129–13141 (2018).
12. Siwicki, B. (2020) Updated: A Guide to Connected Health Device and Remote Patient Monitoring Vendors. Available online: https://www.healthcareitnews.com/news/guide-connected-health-device-and-remote-patient-monitoring-vendors/ (accessed on 21 June 2022).
13. Poppas, A., Rumsfeld, J.S., Wessler, J.D.: Telehealth is having a moment: Will it last? *J. Am. Coll. Cardiol.* 75, 2989–2991 (2020).
14. Suh, M.K., Chen, C.A., Woodbridge, J., Tu, M.K., Kim, J.I., Nahapetian, A., Evangelista, L.S. and Sarrafzadeh, M., 2011. A remote patient monitoring system for congestive heart failure. Journal of medical systems, 35, pp.1165-1179.
15. Gia, T.N., Amir-Mohammad Rahmani, T.W.P.L., Tenhunen, H.: Fault tolerant and scalable IoT-based architecture for health monitoring. *IEEE Access* (2015), pp. 1–6.
16. Rohokale, V. M., Prasad, N.R., Prasad, R.: A cooperative Internet of Things (IoT) for rural healthcare monitoring and control. In: *Proc. Int. Conf. wireless communication, vehicular technology, information theory and aerospace & electronic systems technology (Wireless VITAE),* Chennai, India pp. 1–6 (2011).
17. Chung, W.-Y., Lee, Y.D., Jung, S.J.: A cooperative Internet of Things (IoT) for rural healthcare monitoring and control. In: *A Wireless Sensor Network Compatible Wearable u-Healthcare Monitoring System Using Integrated ECG, Accelerometer and SpO2,* pp. 1529–1532 (2008), Vancouver, Canada.
18. Zhao, C.W., Nakahira, Y.: Medical application on IoT. In: *International Conference on Computer Theory and Applications (ICCTA), Alexandria, Egypt* pp. 660–665 (2011).
19. Fan, Y.J., Yin, Y.H., Xu, L.D., Zeng, Y., Wu, F.: IoT-based smart rehabilitation system. *IEEE Trans. Ind. Informat.* 10, 1568–1577 (2014).
20. Wan, S., Gu, Z., Ni, Q. Cognitive computing and wireless communications on the edge for healthcare service robots. *Comput. Comm.* 149, 99–106 (2020).
21. Baur, K., Schättin, A., de Bruin, E.D., Riener, R., Duarte, J.E., Wolf, P.: Trends in robot-assisted and virtual reality-assisted neuromuscular therapy: A systematic review of health-related multiplayer games. *J. Neuroeng. Rehabil.* 15(1), 107 (2018).
22. Tao, V., Moy, K., Amirfar, V.A.: A little robot with big promise may be future of personalized health care. *Pharmacy Today* 22(9), 38 (2016).
23. Ritschel, H., Seiderer, A., Janowski, K., Aslan, I., André, E. Drink-O-Mender: An adaptive robotic drink adviser. In: *Proceedings of the 3rd International Workshop on Multisensory Approaches to Human-Food Interaction. 2018 Presented at: MHFI'18,* October 16, 2018, Boulder, CO.
24. Moyle, W., Jones, C., Cooke, M., O'Dwyer, S., Sung, B., Drummond, S.: Social robots helping people with dementia: Assessing efficacy of social robots in the nursing home environment. In: *6th International Conference on Human System Interactions (HSI). 2013 Presented at: HSI'13,* June 6–8, 2013, Sopot, Poland.
25. Bhelonde, A., Didolkar, N., Jangale, S., Kulkarni, N.: Flexible wound assessment system for diabetic patient using android smartphone. In: *International Conference on Green Computing and Internet of Things. 2015 Presented at: ICGCIoT'15,* October 8–10, 2015, Noida, India.
26. Ashique, K., Kaliyadan, F., Aurangabadkar, S.J.: Clinical photography in dermatology using smartphones: An overview. *Indian Dermatol. Online J.* 6(3), 158–163 (2015).

4 Blockchain-Enabled Data Security and Integrity in IoT-Big Data Systems for Smart Cities

*Tarun Kumar Vashishth, Vikas Sharma,
Kewal Krishan Sharma, Bhupendra Kumar,
Sachin Chaudhary, and Rajneesh Panwar*

4.1 INTRODUCTION

In the contemporary landscape of smart cities, the exponential growth of Internet of Things (IoT) instruments has ushered in an era of unparalleled data generation, raising profound concerns about the security and integrity of massive volumes of information produced. This chapter embarks on a comprehensive exploration of a transformative paradigm, centering on the integration of blockchain technology to fortify data security and guarantee the integrity of IoT-generated big data within smart city environments. As the proliferation of interconnected devices continues, traditional security frameworks face challenges in addressing vulnerabilities associated with data breaches and tampering. Leveraging blockchain's intrinsic features, including decentralization, immutability, and cryptographic hashing, emerges as a robust solution to these challenges. By delving into the amalgamation of blockchain into IoT-Big Data systems, this study aims to elucidate the establishment of a secure and transparent data management ecosystem, ensuring the safeguarding of data authenticity and lineage throughout its lifecycle.

4.1.1 BACKGROUND

The evolution of smart cities in the digital era has been propelled by the pervasive deployment of IoT devices, fostering an unprecedented surge in data generation. As these urban landscapes become increasingly interconnected, the ensuing deluge of data brings forth profound challenges, most notably in the realms of data security and integrity. Traditional data management systems grapple with the complexities arising from the sheer volume and heterogeneity of IoT-generated big data. This chapter delves into the forefront of technological innovation by exploring the integration of blockchain—a distributed ledger technology—as a formidable solution to the burgeoning concerns surrounding data in smart cities. Inherent features of blockchain,

DOI: 10.1201/9781032673479-5

including decentralization, immutability, and cryptographic hashing, present a compelling framework for improving security posture and preserving the integrity of IoT-generated data. In light of the escalating threats posed by data breaches and tampering, blockchain offers a paradigm shift, promising transparency, accountability, and resilience. By elucidating the symbiotic relationship between blockchain and IoT-Big Data systems, this chapter goals to contribute to a nuanced thoughtful of how this integration can foster a secure, transparent, and trustworthy data management ecosystem, laying the foundation for resilient and future-ready smart cities.

4.1.2 Objectives of the Chapter

The objectives of this chapter are multifaceted, driven by the imperative to address the critical challenges posed by the propagation of IoT devices and the consequential surge in data volumes within the context of smart cities. Foremost, the chapter seeks to illuminate the pressing concerns surrounding data security and integrity in smart cities, examining the vulnerabilities that emerge with the relentless growth of IoT-generated big data. By delving into the landscape of traditional data management systems and their limitations in coping with the dynamic nature of smart city environments, the chapter aims to underscore the necessity for innovative solutions. The primary goal is to explore the transformative potential of blockchain technology in mitigating these challenges, offering a decentralized, immutable, and cryptographically secure framework for bolstering data security and ensuring the integrity of information flows. Furthermore, the chapter endeavors to articulate the specific objectives related to the integration of blockchain into IoT-Big Data systems, elucidating how this integration can establish a secure and transparent data management ecosystem. The study seeks to dissect various blockchain-enabled security mechanisms, such as access control, data provenance, and consensus algorithms, with the overarching aim of providing a comprehensive understanding of how blockchain can be harnessed to fortify data protection in the intricate and interconnected landscapes of smart cities.

4.1.3 Significance of Blockchain in IoT-Big Data Systems for Smart Cities

Significance of integrating blockchain into IoT-Big Information systems for cities is paramount, stemming from its transformative potential to address critical challenges in data security and integrity. As smart cities increasingly rely on the IoT for data-driven decision-making, the sheer volume and heterogeneity of generated data introduce vulnerabilities that traditional systems struggle to mitigate. Blockchain's inherent characteristics, including decentralization, immutability, and cryptographic hashing, offer a paradigm shift in securing this complex data ecosystem. By providing a decentralized ledger, blockchain ensures that data is not concentrated in a single point of control, lowering the risk of non-authorized access or tampering. Immutability of blockchain records guarantees the integrity of data, establishing a transparent and traceable lineage for all information generated by IoT devices. This is particularly crucial in smart city contexts where the reliability of data is paramount

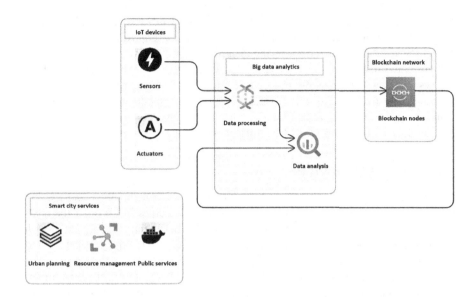

FIGURE 4.1   Blockchain and IoT-Big Data systems integration for smart cities.

for urban planning, resource management, and public services. Furthermore, block-chain's cryptographic mechanisms enhance data confidentiality, authenticity, and access control. The significance of this integration lies in fortifying the foundation of smart cities, instilling trust in the data generated by IoT devices with fostering a resilient, and efficient information management ecosystem essential for the sustainable development of urban environments (Figure 4.1).

4.1.4   STRUCTURE OF THE CHAPTER

The chapter unfolds systematically to comprehensively address the challenges of data security and integrity in IoT-Big Data systems within smart cities. It commences with an Introduction, laying the groundwork by highlighting the surge in IoT devices and resultant data volumes in smart urban environments. Following this, Section 4.1.1 provides an insightful examination of the current state of IoT-Big Data systems, elucidating the challenges and vulnerabilities inherent in their frameworks. Section 4.1.2 explicitly outlines the goals of the research, emphasizing the need to explore blockchain's transformative potential in addressing these challenges. Subsequently, the Significance of Blockchain in IoT-Big Data Systems for Smart Cities under-scores the pivotal role of blockchain in fortifying data security, ensuring integrity, and establishing a secure data management ecosystem. The chapter then navigates through the technical terrain of Blockchain Technology and Its Relevance to IoT-Big Data Systems, detailing the features of blockchain and their applicability in smart city contexts.Blockchain integration in IoT-Big Data Systems section delves into the practical aspects, discussing seamless integration strategics, benefits, and potential challenges. The establishment of a secure data management ecosystem is explored, shedding light on how blockchain's distributed ledger and consensus mechanisms contribute to safeguarding IoT-generated data. An in-depth analysis of blockchain-enabled security mechanisms follows, evaluating access control, data provenance, and consensus algorithms for bolstering data protection. Real-world applications are then presented in Section 4.9, illustrating successful implementations of blockchain in cities. Section 4.10 critically examines implementation challenges and proposes avenues for future research. Finally, the conclusion succinctly summarizes the key findings, emphasizing the transformative role of blockchain in mitigating data security and integrity concerns within the intricate landscape of smart cities.

4.2 LITERATURE REVIEW

Abbas et al. (2021) discussed the potential of blockchain technology and the IoT to enhance security of transportation systems in smart cities. It outlined the challenges and opportunities associated with the use of these technologies, along with potential solutions for overcoming them. The authors concluded that blockchain and IoT could be used to improve the security of transportation systems in smart cities, and that more research was desirable to explore the full potential of these technologies. Abd et al. (2021) examined the potential of quantum stimulated blockchain-based cybersecurity for securing smart edge utilities in the context of smart cities. It discusses the challenges and potential solutions for implementing such systems and provides a framework for their development. Ahmed et al. (2022) proposed a design to allow for improved data sharing among stakeholders, enhanced security, reduced energy consumption, and improved management of resources. The authors found that their framework could be used to create more efficient and sustainable cities. Awotunde et al. (2023) discussed the use of hybrid deep learning-enabled blockchain for the privacy and security enhancement of smart cities. The authors explored various techniques to ensure the security of smart city data and proposed a hybrid deep learning-enabled blockchain method to secure the data. They concluded that the proposed method could provide better security than traditional methods. Chinnasamy et al. (2021) discussed the potential of blockchain know-how in smart cities in their article 'Blockchain Technology in Smart-Cities'. They analyzed the various applications of the technology and the challenges faced in its implementation. The authors concluded that blockchain has the power to revolutionize the way cities are managed and operated. El Bekkali et al. (2023) proposed a secure, distributed ledger-based framework that leverages distributed consensus to authenticate and validate the data and transactions in smart cities. The framework is designed to protect against cyber-attacks, malicious transactions, and data manipulation.

Helen (2023) explored cyber outbreaks in blockchain technology enabled green cities. She examined the ways these attacks could exploit blockchain technology and how this technology could be used to protect cities from these risks. Juma et al. (2023) proposed the use of Trusted Consortium Blockchain (TCB) to secure Bigdata integrity for Industrial Internet of Things in Smart Manufacturing. They argued that the TCB approach could offer robust and efficient data integrity protection through the elimination of the single point of failure. Their findings showed that the TCB approach could be used to effectively secure Bigdata integrity for Industrial IoT in Smart Manufacturing. Kant et al. (2023) presented a deployment mechanism for

IoT-based security based on blockchain technology. Khawaja and Javidroozi (2023) discussed the potential of blockchain technology to enable cross-sectoral systems combination for development of sustainable cities. The article highlighted the ability of blockchain technology to facilitate secure and transparent distributed transactions, which could enable the integration of multiple systems to create smart towns. Rahman et al. (2021) discussed a framework to provide secure, private and explainable Internet of Healthcare Things (IoHT) for sustainable health monitoring in a smart city. The authors proposed a framework that addressed the security and privacy issues associated with IoHT while maintaining accountability, explainability and trust. The authors also discussed the challenges and opportunities associated with developing such a framework for sustainable health monitoring in a smart city. Sharma and Kumar (2023) discussed the role of Artificial Intelligence (AI) to enhance the security and privacy of data in smart cities. The article explored the potential of AI to increase the safety of citizens by protecting their data from unauthorized access and manipulation. It also discussed the challenges of implementing AI-based tools in smart cities, such as lack of data privacy regulations, the potential for misuse of AI, and need for secure data collection and storage. The authors concluded that while AI has the power to advance information security and privacy in smart cities, additional research is needed to address challenges associated with its implementation. Shen et al. (2020) presented a special issue on the trust-oriented designs of IoT for smart cities. It provided an overview of the trust-oriented designs of IoT and discussed the challenges and opportunities of the technology for smart cities. Singh et al. (2020) highlighted how the combination of blockchain and AI could be used to improve the security of communication networks, as well as create new opportunities for the development of a sustainable and secure urban environment. The research concluded that blockchain and AI can be effectively used to improve the security, scalability, and privacy of IoT networks and thus facilitate the development of smart cities. Vashishth et al. (2023) discussed the various ways in which AI has been used to improve computing systems and the challenges that it poses. The article concluded that AI offers a promising solution to the problems of current computing systems and will be a transformative force in the future. Yadav et al. (2023) explored the power of blockchain technology in abolishing the prevalent threats in IoT-based applications and smart cities. Through the analysis of current research and challenges, the authors proposed a non-centralized architecture for secure data sharing and access control using blockchain. The implications of the proposed model were tested through a simulated environment. Zhaofeng et al. (2019) proposed a solution to handle security and privacy issues related with the IoT information. The article examines different security protocols and trust management models to ensure the secure data.

4.3 IOT-BIG DATA LANDSCAPE IN SMART CITIES

IoT-Big Data landscape in smart towns is marked by an unprecedented proliferation of interconnected devices generating vast volumes of data. These devices, ranging from sensors and actuators to surveillance cameras and smart infrastructure, contribute to a dynamic and complex urban data ecosystem. The continuous streams of data hold immense potential for informing decision-making processes in areas

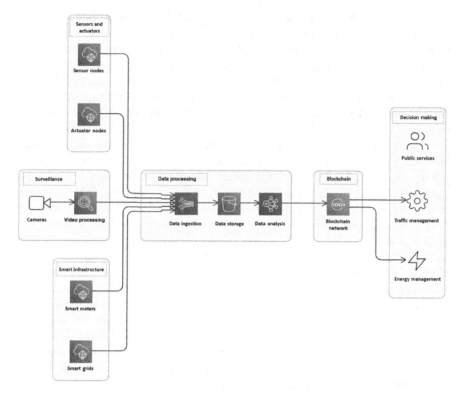

FIGURE 4.2   IoT-Big Data in smart cities.

such as traffic management, energy consumption, and public services. However, the sheer scale and heterogeneity of this data pose substantial challenges in terms of security, integrity, and efficient management. Traditional centralized systems often struggle to cope with the intricacies of real-time data flows, creating vulnerabilities that blockchain technology aims to address by providing a decentralized, transparent, and secure framework for managing IoT-generated big data in smart city environments (Figure 4.2).

4.3.1   PROLIFERATION OF IoT DEVICES

The IoT-Big Data landscape in smart cities is characterized by an unprecedented proliferation of IoT devices, constituting a foundational element of the urban fabric. From smart sensors embedded in infrastructure to connected devices in homes and businesses, a diverse array of IoT sensors creates huge volumes of information. Such devices contribute to the real-time monitoring and collection of information critical for the effective functioning of smart city initiatives. The proliferation encompasses a spectrum of applications, including traffic management, environmental monitoring, healthcare systems, and energy grids. This interconnected network of devices enables cities to gather insights, optimize resource allocation, and enhance overall urban efficiency. However, this proliferation also introduces significant challenges, particularly in managing the sheer magnitude and heterogeneity of data streams. The dynamic nature of smart city environments demands solutions that can secure, authenticate, and maintain the integrity of this continuous influx of data. As a response to such encounters, blockchain technology emerges as a capable solution, provided that a decentralized and secure framework ensures the reliability of IoT-generated big data in smart city ecosystems, thereby contributing to the sustainability and resilience of urban infrastructure.

4.3.2   GENERATION OF MASSIVE DATA IN SMART METROPOLITAN

"Generation of Massive Information in Cities of Blockchain" delves into the substantial data volumes produced within the context of smart urban environments and explores how blockchain technology can offer innovative solutions to manage and secure this wealth of information. In smart cities, the incorporation of IoT devices, sensors, and interconnected infrastructure results in an incessant stream of diverse and extensive datasets. These datasets span various sectors, such as transportation, healthcare, energy, and public services, contributing to the efficiency and optimization of urban operations. However, the massive scale and dynamic nature of this data pose challenges in terms of storage, processing, and maintaining its integrity. The topic investigates the role of blockchain in addressing these challenges by enabling a decentralized and transparent framework. The decentralized nature of blockchain, combined with features like immutability and cryptographic security, is explored for its potential to enhance the management, authenticity, and integrity of the large-scale data flow characteristic of smart cities. This exploration contributes to understanding how blockchain technology can be harnessed to fortify data management practices in smart cities, ensuring resilience, security, and efficiency in the face of burgeoning data generation.

4.3.3   SERIOUS CONCERNS: DATA SECURITY AND INTEGRITY

Critical concerns of data security and integrity represent foundational pillars in empire of information management, mainly in the context of evolving technologies with complex data landscapes. Data security involves safeguarding data from unauthorized admission, breaks, or malicious actions, securing integrity. It encompasses protective measures such as encryption, access controls, and authentication protocols to prevent unauthorized entities from compromising sensitive data.

Integrity, on the other hand, pertains to correctness and consistency of information. Maintaining data integrity safeguards that data leftovers untouched and trustworthy throughout its lifecycle. Any unauthorized alterations, corruption, or tampering with data compromises its integrity, potentially leading to misinformation or faulty decision-making.

In contemporary scenarios, where data is constantly generated, transmitted, and processed, these concerns are amplified. Technologies like blockchain have emerged as innovative solutions to address data security and integrity challenges. Decentralized of Blockchain's immutable nature ensures that data, once recorded, cannot be changed, providing a secure and transparent framework. As data continues to play a pivotal role in decision-making processes across various industries, addressing these critical concerns becomes imperative for maintaining trust, reliability, and the effective functioning of information-driven systems.

4.4   FOUNDATIONAL CONCEPTS OF BLOCKCHAIN

4.4.1   Overview of Blockchain Technology

Blockchain technology serves as the bedrock for providing information preservation and integrity in the kingdom of IoT-Big Data systems within cities. At its core, a blockchain is a decentralized and distributed ledger that records transactions across a network of nodes. Every transaction is bundled into a block, and these blocks are linked through cryptographic hashes, forming a chain. This architecture ensures transparency, as all participants in network have a synchronized copy of the entire ledger. The use of consensus mechanisms, such as proof-of-work or proof-of-stake, validates the transactions, making it computationally infeasible for a single entity to manipulate the entire blockchain. This decentralized and transparent structure not only improves the security of data but also provides a trustworthy framework for managing vast amounts of data created by devices of IoT in intricate landscape of smart cities (Figure 4.3).

4.4.2   Decentralization in Blockchain

Decentralization is a foundational concept in blockchain technology, and its significance in the context of IoT-Big Data systems for cities cannot be overstated. In non-centralized blockchain network, there is no reliance on a central authority for data control or validation. Instead, multiple nodes (participants) in network keep copies of blockchain, and every node participates in the validation process. This architecture

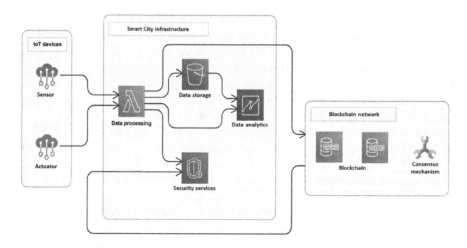

FIGURE 4.3   Significance of blockchain in smart cities.

mitigates the risks related to one point of failure or malicious manipulation. In smart cities, where the proliferation of IoT devices requires a robust and resilient data infrastructure, decentralization ensures that the system is not vulnerable to centralized attacks or unauthorized interference. It fosters a distributed and collaborative approach to data management, aligning seamlessly with the dynamic and interconnected nature of smart city environments.

4.4.3 IMMUTABILITY AND CRYPTOGRAPHIC HASHING

Immutability, coupled with cryptographic hashing, forms the backbone of blockchain's ability to guarantee the integrity of data. Immutability ensures that once a block of data is added to blockchain, it becomes tamper-resistant and cannot be changed retroactively. It is achieved through use of cryptographic hashes, which are unique and irreversible representations of data. Individually block contains a hash of the preceding block, forming a continuous chain. Every attempt to change data in a block would require changing its hash, which would demand altering the hashes of entire succeeding blocks. The computational complexity of such an endeavor makes the blockchain highly secure against tampering. In the context of IoT-generated big data in cities, this immutability guarantees the authenticity of information. As data flows from myriad IoT devices, blockchain's cryptographic integrity ensures that the information remains unaltered and reliable, providing a foundation for trusted decision-making processes.

4.4.4 PERTINENCE OF BLOCKCHAIN TO SMART CITY ENVIRONMENTS

The applicability of blockchain to smart city environments stems from its unique features and addresses the specific challenges posed by the dynamic and data-intensive nature of urban landscapes. Blockchain's decentralized architecture aligns with the distributed nature of smart cities, where diverse entities and systems need to collaboratively manage and access data. In the realm of IoT-generated big data, blockchain ensures the integrity of information by providing a tamper-resistant framework. As smart cities rely on data for crucial decision-making processes, especially in areas such as management of traffic, optimization of energy, and public services, blockchain's ability to create an unforgeable and transparent ledger is invaluable. The immutability of blockchain guarantees that the data generated by IoT devices, whether from traffic sensors, environmental monitors, or smart infrastructure, remains unchanged and trustworthy throughout its lifecycle. By addressing concerns related to centralized vulnerabilities and data tampering, blockchain emerges as a transformative solution for establishing secure and resilient data management ecosystems in the complex milieu of smart city environments.

4.5 ADDRESSING VULNERABILITIES WITH BLOCKCHAIN

In paradigm of IoT-Big Data systems for good cities, addressing vulnerabilities is paramount, and blockchain technology plays a pivotal role in fortifying data security and integrity. The decentralized and transparent nature of blockchain fundamentally transforms how vulnerabilities are approached. One critical aspect is the prevention of data breaches, and this is explored in detail in subsequent sections. Moreover, blockchain introduces a distributed and consensus-based approach that significantly reduces the risk of unauthorized access or manipulation. By distributing data across a network of nodes and requiring consensus for any changes, blockchain creates a resilient barrier against centralized attacks. This decentralized architecture ensures that compromising a single point in the network doesn't compromise the entire system, thereby elevating the security posture of IoT-generated big data in cities. This section delves into the overarching theme of how blockchain addresses vulnerabilities comprehensively, establishing a foundation for robust data security protocols in the dynamic and interconnected smart city environments.

4.5.1 Role of Blockchain in Data Breach Prevention

Data breach prevention stands as a critical imperative in era of IoT-generated Bigdata, especially within intricate landscapes for smart cities. Blockchain's role in preventing data breaches is transformative, primarily due to its decentralized architecture and cryptographic security features. Unlike traditional systems vulnerable to centralized breaches, blockchain distributes data across a network of nodes, eliminating a single point of failure. The use of cryptographic hashing ensures that sensitive information is securely stored and transmitted. Additionally, blockchain's consensus mechanisms, whether proof-of-work or proof-of-stake, require majority agreement for any changes to the data, making unauthorized access or manipulation exceedingly difficult. This section explores the mechanics of how blockchain functions as a proactive deterrent to data breaches, establishing a resilient and tamper-resistant environment for the diverse streams of data flowing from IoT devices in smart cities. As cities increasingly rely on data-driven decision-making, the prevention of data breaches becomes a cornerstone of ensuring the reliability and security of information.

4.5.2 Safeguarding against Data Tampering

Safeguarding against data tampering is a critical aspect of ensuring the integrity of IoT-generated big data in smart cities, and blockchain stands as a robust solution to this challenge. The immutability inherent in blockchain technology ensures that once data is recorded on the ledger, it cannot be altered retroactively. Every block holds a cryptographic hash of the previous block, creating a chain that is resistant to tampering. Attempting to alter the information in a single block would require changing its hash, which in turn would necessitate altering hashes of all subsequent blocks, a computationally infeasible task. This section elucidates the mechanics of how blockchain's cryptographic security features, coupled with its decentralized and distributed architecture, create an environment where data tampering becomes practically impossible. As smart cities heavily rely on accurate and unaltered data for urban planning, resource optimization, and public services, safeguarding against data tampering becomes integral to ensuring the reliability and authenticity of information.

4.5.3 LEVERAGING BLOCKCHAIN FOR ENHANCED DATA SECURITY

Leveraging blockchain for enhanced data security encompasses a holistic approach that goes beyond preventing breaches and tampering. Blockchain's decentralized nature ensures that data is not stored in a central repository vulnerable to attacks, and cryptographic mechanisms provide an additional layer of protection. This section explores how blockchain can be leveraged to establish secure data management ecosystems in smart cities. It delves into the role of access control mechanisms in blockchain, ensuring that only authorized entities can interact with and validate information. Additionally, the transparency introduced by blockchain allows for improved traceability of data provenance, offering insights into the origin and journey of information. Consensus algorithms further enhance data security by requiring majority agreement before changes are accepted. The section navigates through these various aspects, showcasing how blockchain serves as a multifaceted tool for elevating data security standards in the dynamic and interconnected landscapes of smart cities, where the reliability of data is paramount.

4.6 BLOCKCHAIN INTEGRATION IN IOT-BIG DATA SYSTEMS

Integration of blockchain into IoT-Big Data systems represents a transformative step toward enhancing data safety and integrity in cities. This section explores the seamless integration strategies that enable the harmonious coexistence of these technologies. Seamless integration involves aligning the decentralized, transparent nature of blockchain with the dynamic and interconnected landscape of IoT-generated big data. By providing a distributed ledger for recording and validating transactions, blockchain becomes an integral part of the data management ecosystem. This section delves into the technical considerations and methodologies that facilitate the smooth amalgamation of blockchain into existing IoT-Big Data architectures, fostering a holistic approach to secure and transparent information management in the complex environments of smart cities.

4.6.1 SEAMLESS INTEGRATION STRATEGIES

Seamless integration of blockchain in IoT-Big Data systems is essential for optimizing their performance. The following strategies can be employed to achieve seamless integration:

Integration at the Edge: This approach involves implementing blockchain directly at the edge devices or gateways, which interact with IoT sensors. This integration strategy minimizes latency and maximizes security, as the blockchain network is distributed across the edge devices.

Middleware Approach: A middleware approach can be utilized to connect IoT-Big Data systems with blockchain networks. Middleware enables communication between the IoT-Big Data systems and the blockchain network by converting data formats, interpreting commands, and ensuring the secure and efficient exchange of data.

Distributed Architecture: Blockchain can be implemented as a distributed architecture in IoT-Big Data systems. This architecture enables the blockchain network to grow dynamically, adapting to the changing requirements of the IoT-Big Data system.

Cloud-Based Blockchain Services: IoT-Big Data systems can be integrated with cloud-based blockchain services. These services offer scalable and efficient solutions for blockchain network management and provide the necessary tools and infrastructure for IoT-Big Data system integration.

Use of Interoperable Standards: Ensuring interoperability between IoT-Big Data systems and blockchain networks is crucial for seamless integration. Various industry-recognized standards, such as Hyperledger Fabric and Ethereum, can be used to facilitate this interoperability.

4.6.2 BENEFITS OF INTEGRATING BLOCKCHAIN

Integrating blockchain in IoT-Big Data systems offers several benefits, including:

Enhanced Security: Blockchain's inherent security features, such as encryption and decentralization, can be employed to protect IoT-Big Data systems from unauthorized access and malicious activities.

Improved Data Integrity: Blockchain's immutable and tamper-proof nature ensures accuracy and data reliability in IoT-Big Data systems. This helps in preventing data inconsistencies and reduces the risk of incorrect decisions based on incomplete or incorrect data.

Transparent Data Sharing: Blockchain allows for transparent and secure data sharing among various stakeholders in IoT-Big Data systems. This enables collaboration and fosters trust among participants.

Enhanced Decision-Making: Blockchain-enabled IoT-Big Data systems can provide real-time, up-to-date data to support faster and more accurate decision-making. This can lead to improved efficiency and cost savings.

Scalability and Efficiency: Blockchain's scalability and ability to handle a large volume of transactions make it suitable for large-scale IoT-Big Data systems. This increased efficiency and costs reduced (Figure 4.4).

4.6.3 CHALLENGES IN INTEGRATING BLOCKCHAIN TECHNOLOGY

Despite the potential benefits, integrating blockchain technology in IoT-Big Data systems presents several challenges:

Scalability Concerns: The high energy consumption and storage requirements of blockchain technology may not be feasible for IoT-Big Data systems with limited resources. Additionally, blockchain's limited scalability can impact its adoption in IoT-Big Data systems.

Interoperability Issues: Ensuring interoperability between IoT-Big Data systems and blockchain networks can be challenging, as different systems may have varying data formats, security protocols, and APIs.

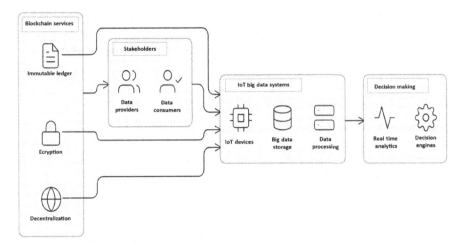

FIGURE 4.4 Benefits of blockchain in smart cities.

High Entry Barriers: Blockchain technology is complex and requires specialized knowledge to implement and maintain. This can present a barrier to entry
for smaller IoT-Big Data systems without the necessary resources or expertise.

Data Privacy Concerns: Ensuring data privacy while maintaining security
and integrity in blockchain-enabled IoT-Big Data systems can be challenging. Existing privacy-enhancing technologies, such as zero-knowledge
proofs, can help mitigate these concerns.

Cost Efficiency: While blockchain offers potential cost savings and efficiency
improvements, its initial implementation costs may be prohibitive for
smaller IoT-Big Data systems with limited budgets.

In conclusion, while integrating blockchain technology in IoT-Big Data systems
presents several challenges, its benefits, like enhanced security, data integrity, and
decision-making capabilities, make it an attractive option for organizations looking
to enhance the performance and security of their IoT-Big Data systems. Addressing
the challenges associated with integration and implementing effective strategies can
help mitigate these risks.

4.7 ESTABLISHING A SECURE DATA MANAGEMENT ECOSYSTEM

To achieve data security and integrity in IoT-Big Data organizations of cities, a robust
data management ecosystem must be established. This ecosystem will ensure that
data is managed securely, consistently, and transparently throughout the IoT-Big
Data system's lifecycle.

4.7.1 UTILIZING BLOCKCHAIN'S DISTRIBUTED LEDGER

The first step in establishing a secure data management ecosystem is to utilize blockchain's distributed ledger technology. This decentralized database ensures data

integrity by preventing unauthorized modifications. In IoT-Big Data systems, blockchain's distributed ledger can be employed to store data related to devices, sensors, and applications, providing a secure and tamper-proof platform for managing IoT-Big Data system data.

4.7.2 Consensus Mechanisms for Authenticity and Lineage

In addition to blockchain's distributed ledger, consensus mechanisms are essential for maintaining data authenticity and traceability in IoT-Big Data systems. Consensus mechanisms determine the validity of data and ensure that all participants in the IoT-Big Data system agree on the authenticity and integrity of information.

4.7.3 Transparency in Data Management

To enhance data security and integrity in IoT-Big Data systems, blockchain-based solutions must ensure transparency in data management. This can be achieved by providing a clear audit trail of data transactions and providing easy access to data for authorized users. By implementing transparency, IoT-Big Data systems can prevent data misuse and maintain trust among participants.

4.7.4 Smart Contracts for Automating Data-Driven Decisions

Another essential component of a secure data management ecosystem is the execution of smart contracts. Smart contracts are self-executing contracts with the terms of the contract nonstop written into code. In IoT-Big Data systems, smart contracts can be employed to auto-decision processes built on real-time data. This can help ensure that IoT-Big Data systems can adapt to changing conditions and make data-driven decisions efficiently.

4.7.5 Integration with IoT-Big Data System Components

To ensure the successful integration of blockchain-based data security and integrity solutions into IoT-Big Data systems, these solutions must be integrated with other components of the IoT-Big Data system. This may involve integrating with IoT devices, data storage systems, and data processing platforms.

4.7.6 Scalability and Adaptability

The scalability and adaptability of the data management ecosystem are crucial for ensuring its long-term success in IoT-Big Data systems. This requires continuous improvement of blockchain-based solutions and adapting them to changing IoT-Big Data system requirements. Additionally, considering factors such as energy consumption, data storage, and computational resources may help enhance the scalability and adaptability of the data management ecosystem.

In conclusion, establishing a secure data management ecosystem in IoT-Big Data systems is essential for achieving data security and integrity. By leveraging blockchain's distributed ledger, consensus mechanisms, transparency, smart contracts, integration with IoT-Big Data system components, and ensuring scalability and adaptability, a robust and secure data management ecosystem can be established for smart cities.

4.8 BLOCKCHAIN-ENABLED SECURITY MECHANISMS

4.8.1 ACCESS CONTROL IN BLOCKCHAIN SYSTEMS

In a blockchain-enabled IoT-Big Data system, access control mechanisms play a vital role in maintaining data security and integrity. These mechanisms ensure that only authorized users can access, modify, or delete data. In a blockchain-based IoT-Big Data system, access control can be implemented through various methods, like:

Role-Based-Access-Control: This assigns specific roles to users, each with a different set of permissions. For example, in a smart city's IoT-Big Data system, there might be roles for data scientists, administrators, and application developers, each with specific access permissions (Figure 4.5).

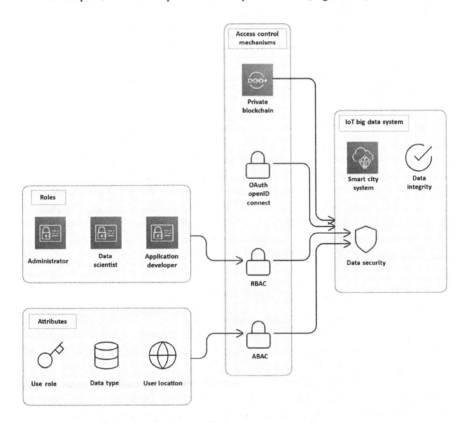

FIGURE 4.5 Access control blockchain IoT data systems.

Attribute-Based-Access-Control (ABAC): ABAC permits more fine-grained admission control by considering the attributes of a user, the accessed data, and the system's environment. For instance, in a smart city's IoT-Big Data system, access to certain data types (e.g., temperature data) might be restricted based on a user's geographical location or their role in the organization.

OAuth and OpenID Connect: These are widely used standards for access control in distributed systems. They allow for secure and flexible access to data and services across different platforms and applications.

Private/Permissioned Blockchain: By using a private or permissioned blockchain, organizations can restrict access to information and the blockchain's network, ensuring that only approved users can access and modify data.

Implementing access control methods in a blockchain-based IoT-Big Data system may help prevent not authorized access to sensitive data, lessen hazard of data breaches, and maintain the overall integrity and safety of the system.

4.8.2 ENSURING DATA PROVENANCE

In IoT-Big Data systems, data provenance refers to the process of tracking the origins, movements, and transformations of data throughout the system. Ensuring data provenance is crucial for maintaining data integrity and trust in IoT-Big Data systems, particularly in applications where the accurate attribution of data is critical. To ensure data provenance in a blockchain-based IoT-Big Data system, data provenance metadata can be stored on the blockchain. This metadata includes information about the data's source, its ownership, its transformation history, and its access control permissions.

For example, in a smart city's IoT-Big Data system, data provenance metadata can be kept on blockchain to track the origins of data collected from sensors installed throughout the city. This information can be used to verify the authenticity and accuracy of the data, as well as to enforce data privacy and access control policies. Implementing data provenance tracking in a blockchain-based IoT-Big Data system can help establishments preserve integrity and trustworthiness of their data, ultimately contributing to success of smart city applications and services.

4.8.3 CONSENSUS ALGORITHMS FOR DATA PROTECTION

In a blockchain-based IoT-Big Data system, consensus algorithms play a critical role in ensuring data security and integrity. These algorithms are responsible for maintaining the integrity of data by verifying and agreeing on the authenticity and consistency of the data. To protect data in a blockchain-based IoT-Big Data system, organizations can implement various consensus algorithms, such as Proof-of-Work, Practical Byzantine Fault Tolerance (PBFT), and Proof-of-Stake. Each consensus algorithm has its own strengths and weaknesses, and the choice of algorithm depends on precise requirements and constraints of IoT-Big Data system.

For example, in a smart city's IoT-Big Data system, organizations can consider using a consensus algorithm that offers a balance between computational resources and energy consumption, such as the Difficulty Adjustment Algorithm used by Bitcoin.

By implementing robust consensus algorithms in a blockchain-based IoT-Big Data arrangement, organizations can help protect integrity and security of their information, ultimately contributing to the success of smart city applications and services.

In conclusion, implementing a comprehensive set of security mechanisms, including access control, data provenance tracking, and consensus algorithms, in a blockchain-based IoT-Big Data system for cities can help ensure data security and integrity. By doing so, organizations can help build trust among stakeholders and facilitate the successful implementation of smart city applications and services.

4.9 CASE STUDIES AND APPLICATIONS

4.9.1 REAL-WORLD EXECUTIONS OF BLOCKCHAIN IN SMART TOWNS

In recent years, there have been several successful real-world implementations of blockchain technology in smart towns. Here are a few examples:

a. Bytom, a decentralized cryptocurrency project, has partnered with various smart city projects to develop solutions for energy management, grid computing, and smart contracts.
b. Singapore has used blockchain technology to manage its national digital identity card, which delivers an efficient way for citizens to authenticate their identity and access digital services.
c. Dubai has been experimenting with practice of blockchain technology in the chain of supply and transportation sectors to enhance security, traceability, and transparency.
d. Microsoft's IoT-Blockchain initiative aims to improve security of IoT systems in smart cities by utilizing blockchain technology.
e. In Brazil, a project called Aetos aims to advance a blockchain-based solution for the management of waste in smart cities.
f. IOTA, a distributed ledger technology, has been used in various smart city applications to enhance the security and scalability of IoT data management.

These examples demonstrate the potential of blockchain in transforming extreme city infrastructure and enhancing the overall quality of life for citizens.

4.9.2 SUCCESSFUL APPLICATIONS AND LESSONS LEARNED

Blockchain technology has power to revolutionize smart city infrastructure and provide numerous benefits, including improved security, efficiency, and transparency. However, it is critical to consider specific necessities and constraints of each smart city project before implementing blockchain solutions.

Here are some lessons learned from successful applications of blockchain know-how in smart cities:

a. **Thorough Understanding of Blockchain Technology**: It is crucial to have a deep understanding of blockchain philosophies, including consensus algorithms, data integrity, and data provenance, to effectively implement and utilize blockchain solutions in smart cities.

b. **Consensus and Collaboration**: To successfully implement blockchain solutions in smart cities, it is necessary to achieve consensus among stakeholders, including governments, private organizations, and citizens. Additionally, collaboration among various entities is essential to ensure the smooth integration and interoperability of different blockchain components.

c. **Regulatory Framework**: For full harness the power of blockchain methodology in smart cities, it is necessary to establish a comprehensive regulatory framework that addresses data security, privacy, and compliance issues.

d. **Customization and Scalability**: To meet the specific needs of smart city projects, blockchain solutions must be customized and designed to scale efficiently, both in terms of data storage and processing capacity.

e. **Data Integration and Interoperability**: To maximize the value of blockchain solutions in sustainable cities, it is important to ensure seamless data integration and interoperability between various systems and platforms, including existing IT infrastructure, IoT devices, and other blockchain networks.

In conclusion, blockchain technology holds immense potential for transforming smart city infrastructure and improving the overall quality of life for citizens. It is essential to carefully consider the specific necessities and restraints of each smart city project and develop comprehensive strategies for the successful implementation and utilization of blockchain solutions.

4.10   CHALLENGES AND FUTURE DIRECTIONS

Future Directions and issues of Blockchain-Enabled Data Security and Integrity in IoT-Big Data Systems for perfect sustainable cities.

4.10.1   CHALLENGES

Scalability: IoT devices and sensors increases in smart cities, ability to scale blockchain-based data security and integrity systems becomes increasingly crucial. While existing blockchain platforms like Ethereum and Hyperledger Fabric have demonstrated their scalability capabilities, the development of efficient and scalable consensus mechanisms remains a significant challenge.

Latency: One of the major challenges of blockchain-enabled data integrity systems in perfect cities is the relatively high latency associated with transaction validation and data propagation. This latency can limit the real-time capabilities of IoT-Big Data systems, especially in critical applications such as smart grid management and

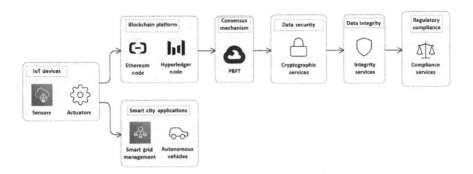

FIGURE 4.6 Challenges in blockchain-enabled data security and IoT in cities.

autonomous vehicles. To overcome this challenge, it is essential to explore advanced consensus mechanisms, such as PBFT, that can minimize latency and transaction time (Figure 4.6).

Data Security and Privacy: As IoT-Big Data systems become increasingly prevalent in smart cities, ensuring privacy and security of sensitive information becomes increasingly important. Blockchain technology has the power to enhance Data Security and Privacy by leveraging cryptographic techniques, such as digital signatures and encryption, to protect sensitive data. However, the successful implementation of privacy-preserving mechanisms in blockchain-enabled IoT-Big Data systems requires careful consideration of various data security challenges, including unauthorized access, data breaches, and malicious tampering.

Data Integrity and Consistency: Power of blockchain technology in ensuring integrity and consistency in IoT-Big Data organizations is significant. However, it is crucial to develop and implement robust mechanisms for maintaining data integrity and consistency in real-time and face-changing scenarios, such as frequent updates and modifications to the underlying blockchain.

Regulatory Compliance: Ensuring regulatory compliance for IoT-Big Data systems using blockchain technology is another important challenge. The successful implementation of blockchain-enabled IoT-Big Data systems requires close collaboration with regulatory authorities to develop and implement compliance frameworks that address data security, privacy, and other regulatory requirements.

4.10.2 Future Directions

Decentralized Autonomous Organizations (DAOs): DAOs, which are self-governing blockchain entities, have the potential to improve the decentralization and safety of IoT-Big Data systems in smart cities. By empowering citizens to directly participate in the governance and management of their city's IoT-Big Data systems, DAOs can promote transparency, accountability, and efficiency.

Smart Contracts and Microservices: To address the challenges associated with latency and scalability, it is significant to explore the development of contract smartness and microservices in blockchain-enabled IoT-Big Data systems.

These innovative solutions can help minimize latency, optimize resource utilization, and enable seamless data integration and interoperability across different IoT devices, sensors, and applications.

Adoption of Federated Blockchain: Federated blockchain, a decentralized consensus mechanism that combines the advantages of blockchain technology with those of traditional federated databases, can potentially offer a more efficient and scalable solution for IoT-Big Data systems in smart cities. By allowing for faster transaction validation and improved data consistency, federated blockchain can help address the challenges associated with traditional blockchain platforms.

Machine Learning and AI Integration: To further enhance the security, efficiency, and scalability of IoT-Big Data systems in smart cities, it is essential to explore the integration of machine learning and AI technologies with blockchain technology. These advanced techniques can help identify potential threats, detect anomalies, and improve overall performance and adaptability of IoT-Big Data systems.

Open Source Development: By promoting open source development and collaboration among developers, researchers, and stakeholders, the adoption of blockchain technology in IoT-Big Data systems for smart cities can be accelerated and made more efficient. Additionally, open source development can help ensure the transparency, security, and interoperability of blockchain-enabled IoT-Big Data systems.

4.11 CONCLUSION

Despite the challenges, integration of blockchain technology in IoT-Big Data systems for smart cities has the potential to transform the way data is secured and utilized. By overcoming obstacles such as scalability, latency, privacy, and data consistency, we can create a more secure, efficient, and interoperable smart city ecosystem. The adoption of DAOs can help promote transparency, accountability, and efficiency in smart city IoT-Big Data systems. DAOs allow citizens to directly participate in the governance and management of their city's IoT-Big Data systems, ensuring that decision-making is decentralized and inclusive. The development of smart contracts and microservices can address the challenges associated with latency and scalability. Smart contracts are self-executing contracts with the terms of the contract directly written into code. By allowing for faster transaction validation and improved data consistency, smart contracts can improve overall security and efficiency of IoT-Big Data systems in smart towns.

Federated blockchain, a decentralized consensus mechanism that combines the advantages of blockchain technology with those of traditional federated databases, can potentially offer a more efficient and scalable solution for IoT-Big Data systems in smart cities. By allowing for faster transaction validation and improved data consistency, federated blockchain can help address the challenges associated with traditional blockchain platforms. Integrating ML and AI technologies with blockchain technology can enhance the security, efficiency, and scalability of IoT-Big Data systems in smart cities. These advanced techniques can help identify potential threats, detect anomalies, and improve the overall performance and adaptability of IoT-Big Data systems. Promoting open source development and collaboration among developers, researchers, and stakeholders can accelerate the adoption of blockchain technology in IoT-Big Data systems for smart towns. By fostering a culture of open collaboration and innovation, open source development can help ensure the transparency, security, and interoperability of blockchain-enabled IoT-Big Data systems.

In conclusion, while challenges related to realizing blockchain technology in IoT-Big Data systems for smart cities are considerable, the potential benefits are undeniable. By overcoming these challenges and continuing to invest in the development of innovative solutions, we can create a more secure, efficient, and interoperable smart city ecosystem. It is crucial to acknowledge that blockchain knowledge is still in its infancy, and as with any emerging technology, its full potential is yet to be realized. However, with ongoing research, collaboration, and innovation, the future of blockchain technology in IoT-Big Data systems for smart cities appears promising. Advancements made in recent years have shown that it is possible to develop secure, efficient, and scalable blockchain-enabled IoT-Big Data systems for smart cities. By continuing to invest in the research and development of these systems, we can harness the power of technology of blockchain to enhance the security, efficiency, and interoperability of smart city infrastructure and expand overall quality of life for its city persons.

REFERENCES

Abbas, K., Lo'Ai, A. T., Rafiq, A., Muthanna, A., Elgendy, I. A., & Abd El-Latif, A. A. (2021). Research article convergence of blockchain and IoT for secure transportation systems in smart cities. *Security and Communication Networks*, 2021(1), 5597679 https://doi.org/10.1155/2021/5597679.

Abd El-Latif, A. A., Abd-El-Atty, B., Mehmood, I., Muhammad, K., Venegas-Andraca, S. E., & Peng, J. (2021). Quantum-inspired blockchain-based cybersecurity: Securing smart edge utilities in IoT-based smart cities. *Information Processing & Management*, 58(4), 102549. https://doi.org/10.1016/j.ipm.2021.102549.

Ahmed, I., Zhang, Y., Jeon, G., Lin, W., Khosravi, M. R., & Qi, L. (2022). A blockchain-and artificial intelligence-enabled smart IoT framework for sustainable city. *International Journal of Intelligent Systems*, 37(9), 6493–6507. https://doi.org/10.1002/int.22852.

Awotunde, J. B., Gaber, T., Prasad, L. N., Folorunso, S. O., & Lalitha, V. L. (2023). Privacy and security enhancement of smart cities using hybrid deep learning-enabled blockchain. *Scalable Computing: Practice and Experience*, 24(3), 561–584.

Chinnasamy, P., Vinothini, C., Arun Kumar, S., Allwyn Sundarraj, A., Annlin Jeba, S. V., & Praveena, V. (2021). Blockchain technology in smart-cities. In Sandeep Kumar Panda, Ajay Kumar Jena, Santosh Kumar Swain, Suresh Chandra Satapathy (Eds.) *Blockchain Technology: Applications and Challenges* (pp. 179–200). Cham: Springer International Publishing. https://doi.org/10.1007/978-3-030-69395-4_11.

El Bekkali, A., Essaaidi, M., &Boulmalf, M. (2023). A blockchain-based architecture and framework for cybersecure smart cities, 11, 76359. *IEEE Access*. https://doi.org/10.1109/ACCESS.2023.3296482.

Helen, D. (2023). Exploring cyber attacks in blockchain technology enabled green smart city. In Saravanan Krishnan, Raghvendra Kumar and Valentina Emilia Balas (Eds) *Green Blockchain Technology for Sustainable Smart Cities* (pp. 343–359). Elsevier, Amsterdam, Netherlands. https://doi.org/10.1016/B978-0-323-95407-5.00005-0.

Juma, M., Alattar, F., &Touqan, B. (2023). Securing big data integrity for industrial IoT in smart manufacturing based on the trusted consortium blockchain (TCB). *IoT*, 4(1), 27–55. https://doi.org/10.3390/iot4010002.

Kant, R., Sharma, S., Vikas, V., Chaudhary, S., Jain, A. K., & Sharma, K. K. (2023). Blockchain-A deployment mechanism for IoT based security. In *2023 International Conference on Computational Intelligence, Communication Technology and Networking (CICTN)* (pp. 739–745), Ghaziabad, India. IEEE. https://doi.org/10.1109/CICTN57981.2023.10140715.

Khawaja, S., & Javidroozi, V. (2023). Blockchain technology as an enabler for cross-sectoral systems integration for developing smart sustainable cities. *IET Smart Cities*, 5(3), 151–172. https://doi.org/10.1049/smc2.12059.

Rahman, M. A., Hossain, M. S., Showail, A. J., Alrajeh, N. A., &Alhamid, M. F. (2021). A secure, private, and explainable IoHT framework to support sustainable health monitoring in a smart city. *Sustainable Cities and Society*, 72, 103083. https://doi.org/10.1016/j.scs.2021.103083.

Sharma, V., & Kumar, S. (2023). Role of artificial intelligence (AI) to enhance the security and privacy of data in smart cities. In *2023 3rd International Conference on Advance Computing and Innovative Technologies in Engineering (ICACITE)* (pp. 596–599), Greater Noida, India. IEEE. https://doi.org/10.1109/ICACITE57410.2023.10182455.

Shen, M., Xu, K., Du, X., Reed, M. J., Bhuiyan, M. Z. A., Zhang, L., &Mijumbi, R. (2020). Guest editorial special issue on trust-oriented designs of Internet of Things for smart cities. *IEEE Internet of Things Journal*, 7(5), 3897–3900. https://doi.org/10.1109/JIOT.2020.2982522.

Singh, S., Sharma, P. K., Yoon, B., Shojafar, M., Cho, G. H., & Ra, I. H. (2020). Convergence of blockchain and artificial intelligence in IoT network for the sustainable smart city. *Sustainable Cities and Society*, 63, 102364. https://doi.org/10.1016/j.scs.2020.102364.

Vashishth, T. K., Kumar, B., Sharma, V., Chaudhary, S., Kumar, S., & Sharma, K. K. (2023). The evolution of AI and its transformative effects on computing: A comparative analysis. In B. Mishra (Ed.), *Intelligent Engineering Applications and Applied Sciences for Sustainability* (pp. 425–442), Hershey, Pennsylvania. IGI Global. https://doi.org/10.4018/979-8-3693-0044-2.ch022

Yadav, L., Mitra, M., Kumar, A., Bhushan, B., & Al-Asadi, M. A. (2023). Nullifying the prevalent threats in IoT based applications and smart cities using blockchain technology. In Devendra Kumar Sharma, Rohit Sharma, Gwanggil Jeon, Zdzislaw Polkowski (Eds) *Low Power Architectures for IoT Applications* (pp. 241–261). Singapore: Springer Nature Singapore. https://doi.org/10.1007/978-981-99-0639-0_14.

Zhaofeng, M., Lingyun, W., Xiaochang, W., Zhen, W., & Weizhe, Z. (2019). Blockchain-enabled decentralized trust management and secure usage control of IoT big data. *IEEE Internet of Things Journal*, 7(5), 4000–4015. https://doi.org/10.1109/JIOT.2019.2960526.

Section 2

The Future of Big Data

5 Impact of Big Data and Internet of Things in Manufacturing
Next Level of Revolution

Ritu Dewan, Arun Kumar Rana, Tapsi Nagpal,
Sumit Kumar Rana, and Shikha Singh

5.1 INTRODUCTION

Internet of Things (IoT) is an ecosystem of interconnected, individually identifiable hardware and software components that can interconnect with one another across a network with little assistance from humans. IoT is described as "a global infrastructure for the information society, enabling advanced services by interconnecting (physical and virtual) things based on existing and evolving interoperable information and communication technologies" [1] by the Internet of Things Global Standards Initiative (IoT-GSI). IoT can be thought of as a ubiquitous, constant connection between the digital and physical worlds [2]. IoT technologies allow physical objects to be remotely controlled and monitored across a network. When coupled with the broad networking of heterogeneous devices, this capacity opens up new possibilities. The physical things that can be connected can be of any scale, from massive automated structures and machinery to tiny embedded electronics in objects, and everything in between. IoT may be imagined as a persistent association between the physical and computerized universes that is all-pervasive [3-6]. IoT innovations empower the further checking and control of physical things over an arrange. This capacity makes way for a modern lesson of applications when combined with the broad network of heterogeneous gadgets. The physical objects that can be associated run in measure from little inserted hardware in things to gigantic mechanized structures, machines, and everything in between. Through the interconnection of numerous technologies and the unification of their functions under one roof, IoT has already destroyed every virtual barrier. According to MarketWatch's figures, the worldwide market for manufacturing IoT products is anticipated to have increased from 12.6 billion USD in 2017 to 45.3 billion USD in 2025 [7]. The adoption of cloud services globally and the growing demand for centralized monitoring and infrastructure management are the key causes of this increase, as shown in Figure 5.1.

DOI: 10.1201/9781032673479-7

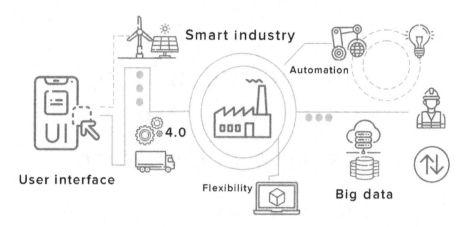

FIGURE 5.1 IoT and Big Data infrastructure management.

Almost no industry operates without the aforementioned technology, whether it be in robots, healthcare, retail, smart cities, or home automation. IoT has a wide range of applications, from telemedicine and remote healthcare to employing public cameras to track social estrangement. IIoT has a significant impact on business operations. This encompasses the methods used in the production of items, client sales, and service afterward. More businesses will begin utilizing IoT in manufacturing and industry as a result of the accessibility of IoT infrastructure and home robots. Now that this has been established, we can affirm that IIoT is one of the main factors underpinning Industry 4.0.

5.2 BACKGROUND AND ENABLING TECHNOLOGIES

There are three fundamental technologies that are fundamental to IoT and have enormous potential for the industrial sector. RFID, or identification by radio frequency, RFID tags attached to objects are automatically recognized and monitored by means of data transmission via electromagnetic fields [9]. RFID systems consist of RFID tags and readers. The objects' unique IDs and other information are stored in RFID tags attached to them. RFID readers that do not need to be in direct line of sight can read these tags and report the information to the business information system. As a result, readers may unintentionally track both the real-time physical movement of the tags and the movement of the objects to which they are attached. Supply chain management, production scheduling, and other uses for RFID are possible in manufacturing. Sensor Networks with Bluetooth The networks known as Wireless Sensor Networks (WSNs) are geographically scattered, autonomous networks of nodes that have the ability to compute, sense their environment, and communicate with one another [10]. The sensor nodes operate independently and decentralized, using multi-hop spreading to send data to the base station while maintaining optimal connectivity for as long as is practical. A single node cannot continuously detect the entire environment; thus, in order to accomplish their tasks, they must cooperate and use collaborative signal and information processing techniques. The nodes may contain actuators that alter the physical characteristics of the world. WSNs have several potential uses in a wide range of sensing-based industrial decision-making situations. Two complementary technologies are WSNs and RFID [11].

RFID can be utilized to recognize and distinguish objects that are difficult to discover or recognize with conventional sensor advances, but it cannot be utilized to screen the state of a protest. WSNs, on the other hand, provide multi-hop remote communication along with protest and environmental state observation. Cloud Computing (CC), which is based on virtualization innovation and empowers compelling administration of an awfully sizable shared pool of reconfigurable computing assets such as systems, servers, and capacity. [12]. On-demand get-to, asset pooling (multi-tenant), fast adaptability, and quantifiable benefits (pay-as-you-go commerce demonstrate) are a few of its key highlights. The fabricating industry may be changed with an offer of assistance from CC [13]. One outstanding endeavor that has picked up around the world consideration is the Cloud Fabricating proposition, an unused service-oriented fabricating worldview [8]. The CC worldview offers unheard-before capabilities for the down-to-earth administration of BD produced by fabricating IoT, much appreciated for its vast computational assets. It all depends on BD whether IoT succeeds or comes up short. BD may be a term for datasets that are as well huge or complicated as ordinary information-preparing procedures [14].

Comparing it to conventional information sets, it has three key "V" characteristics: volume (expansive sums of information, effortlessly bookkeeping for terabytes of information); assortment (heterogeneity of information sorts, organized and unstructured information of content, video, pictures, etc.); and speed (speed of information creation and time outline of information preparing to maximize the esteem) [15]. Others have since proposed including a fourth V (esteem) and a fifth V (veracity) to this list. Stages of information capture, extraction, integration, examination, and elucidation are included within the life expectancy of BD [16]. CC's solid capacity and computation capabilities play a key portion in those stages of movement. The necessities from BD, moreover, rush CC's development [17].

5.3 ARCHITECTURE OF IOT APPLICATIONS IN MANUFACTURING

The SOA comprises services that are discrete and can be combined to accomplish a higher-level objective. The functionality of a physical item may be made available as interoperable services through the use of SOA standards, which are independent of vendors, goods, and technologies. IoT may use SOA because it relies on cooperative communication across heterogeneous devices to accomplish a specified task. The multi-layer SOA's design for IoT in general as well as for particular sectors like manufacturing has been the subject of extensive research [18] and involves three layers: network of devices, CC, and applications. An application layer, a network layer, and a sensor layer make up the three layers of the IoT paradigm [19]. For cyber-physical manufacturing systems, many researchers proposed a five-layer architecture model, with the first layer being the sensing level for data acquisition, the second being the data-to-information conversion level with smart analytics for component health, the third being the cyber level for cyber modeling

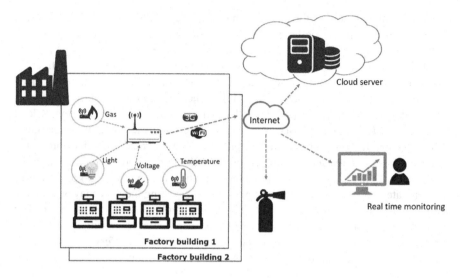

FIGURE 5.2   IoT applications in manufacturing.

and simulation, the fourth being the cognition level for remote visualization, and the fourth being the collaborative-diagnostics and configuration level with self-configuration and self-optimization based on learning [20]. For IoT applications in manufacturing (as shown in Figure 5.2), researchers suggested a real-time information capture and integration system with three layers: event sensing and capturing, manufacturing and data processing, and application services [21].

5.4   IMPACT OF BIG DATA IN MANUFACTURING INDUSTRY

Have you ever wondered how manufacturers use data to enhance their operations and make better decisions? Big Data has the solution. The industrial sector is constantly seeking for methods to improve productivity, reduce costs, and streamline processes. Big Data is also swiftly emerging as a crucial tool for attaining these objectives. Manufacturers may get important insights into their operations, spot opportunities for development, and make wise decisions that can boost efficiency and profitability by gathering and analyzing enormous volumes of data from multiple sources. We will examine seven instances of how Big Data analytics is reshaping the industrial sector in this blog post [22].

Big Data analytics is causing a revolution in the industrial sector. Manufacturers may enhance the quality of their goods, cut costs, optimize their supply chains, and get insights into their manufacturing processes by employing data from multiple sources. Big Data analytics may aid firms in predicting maintenance requirements, avoiding downtime, and fostering a safer workplace. Additionally, manufacturers are now able to make data-driven decisions that can boost growth and profitability thanks to the usage of Big Data in manufacturing. Manufacturers may minimize waste and increase operational efficiency with the correct Big Data strategy.

Big Data analytics may boost growth and profitability while assisting firms to stay ahead of the curve in an increasingly data-driven environment with the correct tools and procedures in place [23–26].

5.5 FUTURE MANUFACTURING APPLICATIONS OF IOT AND BIG DATA

A. **Automation and Efficiency:** IoT and Enormous Information accumulates real-time status data from fabricating floors, counting data on hardware, vehicles, materials, and environment. Without requiring human interest, this information may be utilized to robotize workflows and strategies that keep up and make strides in planning and fabricating frameworks. The control program may naturally make judgments and command actuators to diminish deviations from the arrangement utilizing real-time information collected, intelligent calculations, and organized actuators. Cleverly calculations (like machine learning calculations) with a parcel of multi-source information may naturally give the most excellent The sum of independence, the capacity to manage the generation forms, and the capacity to reply to different unsettling influences has essentially expanded much obliged to progress in machine learning innovation. Large-scale IoT gadget administration (control) requires multi-stakeholder interest and information get to that goes past the conventional solid one-domain and task-specific improvement techniques. Other troubles incorporate the degree of centralization, the taking part in systems' discretionary freedom from one another, and their independent advancement [27–29].

B. **Energy Management:** Vitality administration may be an issue since fabricating accounts for around one-third of the world's vitality utilization [30], and vitality costs are rising. Due to a need for foundation for all-encompassing mapping to commerce and fine-grained, nonstop observing of vitality utilization, conventional strategies are based on discrete plant states without a total picture of the whole plant. By putting sensors at any range of intrigued, IoT and Enormous Information may not as it were persistently degree and relate vitality utilize and trade operations in genuine time, but moreover implement online energetic vitality mindful control within the IoT-enabled "closed" circles. A more comprehensive approach to vitality productivity ought to go past direct stand-alone strategies, such as single process/machine optimization. To create compelling methodologies, cross-domain participation (physical world: apparatus, materials, and vehicles; commerce world: undertaking data frameworks, fabricating forms, and coordinations) is required, as well as information collection and relationship. Also, real-time energy-related records and measurable investigation ought to be joined as an entirety [31].

C. **Proactive Maintenance:** Proactive upkeep, which advances early analysis and portion substitution based on the figure and observing of machine weakening, is, for the most part, recognized by producers as an implication of diminishing costly, spontaneous downtime and unforeseen disappointments [32]. Low-cost sensors, remote associations, and BD devices can give shrewd data about the operation and state of the machine. Machine weakening may be anticipated and seen by modeling, relating, analyzing, and visualizing verifiable and real-time information. Furthermore, for closed circle lifetime re-design, comparative information may be provided back to the item originator.

D. **Connected Supply Chain Management:** Through real-time data, sharing on shop floors, stock, acquiring and deals, support, coordination, etc., IoT-enabled frameworks can interface all parties within the supply chain so that everybody can get it interdependencies, the stream of materials/parts, and generation cycle times, spot potential issues sometime recently they emerge, and take the suitable activity (making a closed control circle). This might have a noteworthy impact on how viably incline fabricating is implemented. Information asymmetry will not exist since all partners will have real-time get to data on request, supply, and input. Common guidelines for information compatibility, information security and security, and financial models for shared data and IoT framework are the essential impediments [33].

5.5.1 IoT and Big Data Challenges in Manufacturing

A. **Security:** More assault vectors are made accessible by IoT and Enormous Information, which increases the number of vulnerabilities that a programmer may take advantage of each step of IoT-driven fabricating, from information collection and handling to further apparatus control, might incorporate security dangers. According to a later Ponemon-established study, information security breaches influence more than 60% of businesses that utilize IoT gadgets [34].

B. **Compatibility:** Because IoT and Big Data platforms and protocols are not standardized, manufacturers have more and more problems connecting their products with other systems and IT infrastructures. One major barrier to manufacturers integrating IoT into their business operations is the lack of regulations around device interaction [35].

   For instance, various businesses use different operating systems (Linux MQTT), programming languages (ARM assembly or JavaScript), and wireless communication protocols (ZigBee, 6LoWPAN, and Thread). This implies that even if you have a collection of items that are able to interact with one another, integrating this method into your organization will cost you a significant amount of time and money [36].

C. **Privacy of Data:** Critical challenges emerge for producers when attempting to get information from sensors built into their items. Producers uncover these gadgets to security perils by interfacing them to the Web. To guarantee adherence to security necessities and the lessening of protection concerns, IoT and enormous data-based fabricating need an intensive arrangement for information assurance from the beginning [37].

Manufacturers are increasingly raising their knowledge of the ways in which IoT and Big Data technology might improve their daily operations in response to these issues. Internet of Things-based solutions are already being used in several industries despite the fact that much work has to be done before this technology can be deemed mature [38].

5.6 CONCLUSION

The primary enabling technologies for IoT and Big Data were examined in this article, including WSNs, smart sensors, Big Data analytics, and CC. A novel four-layer SOA-based design was suggested based on previous architectures for similar systems. Successful IoT and Big Data industrial applications were evaluated and mapped to the suggested four-layer design. The future of competitiveness may be significantly impacted by the integration of smart-sensing and actuation capabilities on equipment with cyberspace, which provides the possibility of new levels of production. A handful of the IoT applications utilized in manufacturing today are cloud-based, but most are not. IoT is widely acknowledged as a unique paradigm with the potential to fundamentally change the industrial sector. It provides the foundation for the development of completely new business and market potential for manufacturing thanks to the highly integrated smart cyber-physical space. Manufacturing that is IoT-enabled is one such initiative that has the potential to have a significant influence on the global economy. This chapter lists many major, ongoing problems with IoT applications in the industrial sector and offers evaluations of some recent developments. Then, several difficult domain-specific difficulties are highlighted together with the prospective IoT industrial applications. In general, new research difficulties brought on by rising IoT applications in manufacturing necessitate technical standards and solutions that can harness massive real-time data streams to improve operations throughout the entire manufacturing life cycle.

REFERENCES

[1] P. Wright, "Cyber-physical product manufacturing," *Manufacturing Letters*, 2014, 2(2): 49–53.
[2] Z. Bi, L. D. Xu, C Wang, "Internet of Things for enterprise systems of modern manufacturing," *IEEE Transactions on Industrial Informatics*, 2014, 10(2): 1537–1546.
[3] L. Atzori, A. Iera, G. Morabito, "The internet of things: A survey," *Computer Networks*, 2010, 54(15): 2787–2805.
[4] D. Miorandi, S. Sicari, F. De Pellegrini, et al., "Internet of things: Vision, applications and research challenges," *Ad Hoc Networks*, 2012, 10(7): 1497–1516.
[5] J. Lee, B. Bagheri, H. A. Kao, "A cyber-physical systems architecture for industry 4.0-based manufacturing systems," *Manufacturing Letters*, 2015, 3: 18–23.
[6] H. Kagermann, J. Helbig, A. Hellinger, et al., "Recommendations for implementing the strategic initiative INDUSTRIE 4.0: Securing the future of German manufacturing industry; final report of the Industrie 4.0 Working Group," *Forschungsunion*, 2013. https://www.din.de/resource/blob/76902/e8cac883f42bf28536e7e8165993f1fd/recommendations-for-implementing-industry-4-0-data.pdf
[7] J. A. Stankovic, "Research directions for the internet of things," *IEEE Internet of Things Journal*, 2014, 1(1): 3–9.

[8] K. Islam, W. Shen, X. Wang, "Wireless sensor network reliability and security in factory automation: A survey," *IEEE Transactions on Systems, Man, and Cybernetics,* 2012, 42(6): 1243–1256.

[9] A. Juels, "RFID security and privacy: A research survey," *IEEE Journal on Selected Areas in Communications,* 2006, 24(2): 381–394.

[10] J. Yick, B. Mukherjee, D. Ghosal, "Wireless sensor network survey," *Computer Networks,* 2008, 52(12): 2292–2330.

[11] L. Wang, L. Da Xu, Z. Bi, et al. "Data cleaning for RFID and WSN integration," IEEE *Transactions on Industrial Informatics,* 2014, 10(1): 408–418.

[12] P. Mell, T. Grance, *Perspectives on Cloud Computing and Standards,* National Institute of Standards and Technology (NIST), Information Technology Laboratory; 2009, Gaithersburg, United States.

[13] B. H. Li, L. Zhang, S. L. Wang, et al., "Cloud manufacturing: A new service-oriented networked manufacturing model," *Computer Integrated Manufacturing Systems,* 2010, 16(1): 1–7.

[14] W. He, L. Xu, "A state-of-the-art survey of cloud manufacturing," *International Journal of Computer Integrated Manufacturing,* 2015, 28(3): 239–250.

[15] V. Schönberger, K. Cukierl, *Big Data: A Revolution That Will Transform How We Live, Work, and Think,* Houghton Mifflin Harcourt, Boston, Massachusetts; 2013.

[16] H. Jagadish, J. Gehrke, A. Labrinidis, et al., "Big data and its technical challenges," *Communications of the ACM,* 2014, 57(7): 86–94.

[17] J. Li, F. Tao, Y. Cheng, et al., "Big data in product lifecycle management," *International Journal of Advanced Manufacturing Technology,* 81, 2015: 1–18.

[18] J. Gubbi, R. Buyya, S. Marusic, et al., "Internet of Things (IoT): A vision, architectural elements, and future directions," *Future Generation Computer Systems,* 2013, 29(7): 1645–1660.

[19] Gupta, B.B. and Quamara, M., 2020. An overview of Internet of Things (IoT): Architectural aspects, challenges, and protocols. *Concurrency and Computation: Practice and Experience, 32*(21), p.e4946.

[20] D. Guinard, V. Trifa, S. Karnouskos, et al., "Interacting with the soabased internet of things: Discovery, query, selection, and on-demand provisioning of web services," *IEEE Transactions on Services Computing,* 2010, 3(3): 223–235.

[21] P. Banerjee, R. Friedrich, C. Bash, et al., "Everything as a service: Powering the new information economy," *Computer,* 2011, 44(3): 36–43.

[22] H. Ning, H. Liu, J. Ma, et al., "Cybermatics: Cyber-physical-social- thinking hyperspace based science and technology," *Future Generation Computer Systems,* 56, 504–522. 2015.

[23] G. G. Meyer, K. Främling, J. Holmström, "Intelligent products: A survey," *Computers in Industry,* 2009, 60(3): 137–148.

[24] S. Veronneau, J. Roy, "RFID benefits, costs, and possibilities: The economical analysis of RFID deployment in a cruise corporation global service supply chain," *International Journal of Production Economics,* 2009, 122(2): 692–702.

[25] W. Shen, Q. Hao, H. Yoon, "Applications of agent-based systems in intelligent manufacturing: An updated review," *Advanced Engineering Informatics,* 2006, 20(4): 415–431.

[26] R. Krishna, A. L. Imoize, R. S. Yaduvanshi, et al., "Analysis of multi-stacked dielectric resonator antenna with its equivalent RLC circuit modeling for wireless communication systems," *Mathematical and Computational Applications,* 2023, 28(1): 4.

[27] S. K. Rana, S. K. Rana, A. K. Rana, et al., A blockchain supported model for secure exchange of land ownership: An innovative approach. *In 2022 International Conference on Computing, Communication, and Intelligent Systems (ICCCIS)* (pp. 484–489). IEEE; Greater Noida, India 2022.

[28] M. Lalit, S.K. Chawla, A. K. Rana, et al., IoT networks: Security vulnerabilities of application layer protocols. In *2022 14th International Conference on Mathematics, Actuarial Science, Computer Science and Statistics (MACS)* (pp. 1–5), Karachi, Pakistan. IEEE; 2022

[29] S. Dhawan, R. Gupta, H. K. Bhuyan, et al., "An efficient steganography technique based on S2OA & DESAE model," *Multimedia Tools and Applications*, 2022: 82, 1–29.

[30] A. K. Rana, S. Sharma, "Internet of Things based stable increased-throughput multi-hop protocol for link efficiency (IoT-SIMPLE) for health monitoring using wireless body area networks," *International Journal of Sensors Wireless Communications and Control*, 2021, 11(7): 789–798.

[31] S. Dhawan, R. Gupta, A. K. Rana, et al., Internet of medical things (IoMT) & secured using steganography for development of smart society 5.0. In Dr. Vikram Bali, Dr. Vishal Bhatnagar, Dr. Joan Lu, Dr. Kakoli Banerjee (Eds) *Decision Analytics for Sustainable Development in Smart Society 5.0* (pp. 173–189). Springer, Singapore; 2022.

[32] A. Kumar, S. Sharma, N. Goyal, et al., "Secure and energy-efficient smart building architecture with emerging technology IoT," *Computer Communications*, 2021, 176: 207–217.

[33] Rana, S.K., Kim, H.C., Pani, S.K., Rana, S.K., Joo, M.I., Rana, A.K. and Aich, S., 2021. Blockchain-based model to improve the performance of the next-generation digital supply chain. *Sustainability*, *13*(18), p.10008..

[34] N. Goyal, S. Sharma, A. K. Rana, et al., editors, *Internet of Things: Robotic and Drone Technology*, CRC Press; 2022, Boca Raton, Florida.

[35] P. Rana, I. Batra, A. Malik, et al., "Intrusion detection systems in cloud computing paradigm: Analysis and overview," *Complexity*, 2022, 2022, 1–14.

[36] R. Gupta, A. K. Rana, S. Dhawan, et al., editors, *Advanced Sensing in Image Processing and IoT*, CRC Press; 2022, Boca Raton, Florida.

[37] S. K. Rana, S. K. Rana, K. Nisar, et al., "Blockchain technology and artificial intelligence based decentralized access control model to enable secure interoperability for healthcare," *Sustainability*, 2022, 14(15): 9471.

[38] F. J. Abdullayeva, "Internet of things-based healthcare system on patient demographic data in Health 4.0," *CAAI Transactions on Intelligence Technology*, 2022, 7(4): 644–657.

6 Application of Artificial Intelligence-Based Digital Asset Management in Healthcare Ecosystem

Lavisha and Mandeep Kaur

6.1 INTRODUCTION

In the evolving healthcare field, using state-of-the-art technology has become crucial to optimize efficiency, boost patient results, and simplify processes. Artificial Intelligence (AI) is a remarkable technology that has been increasingly popular recently. Its use in the healthcare industry, AI, and digital asset management (DAM) is causing a substantial transformation [1]. Integrating AI and digital asset management (DAM) has significant potential to revolutionize the management, accessibility, and use of medical data. This integration can lead to a new era of personalized medicine and data-driven decision-making in healthcare. At its essence, Digital Asset Management encompasses the storage, organization, and retrieval of digital assets, including photographs, videos, documents, and other multimedia files [2].

Advanced systems are required in the medical sector to handle, analyse, and extract valuable insights from the large and complicated volume of medical data created daily. AI enhances DAM systems by introducing cognitive capabilities, resulting in a mutually beneficial partnership that enhances the efficiency and effectiveness of healthcare procedures. Medical imaging is one of the main areas where AI-based Digital Asset Management (DAM) is extensively used in healthcare. Radiology, pathology, and other diagnostic imaging modalities provide vast visual data [3]. When incorporated into digital asset management (DAM) systems, AI algorithms can analyse these photos with unparalleled speed and precision. AI-driven picture identification, for example, can aid in promptly identifying problems, enabling timely treatments and enhancing diagnostic precision. This accelerates healthcare professionals' decision-making process and improves patient outcomes by providing prompt and accurate treatments.

Moreover, the capacity of AI to acquire knowledge and adjust itself over some time renders it an indispensable instrument for categorizing and arranging data within DAM systems. Healthcare organizations need help maintaining many forms of data, such as electronic health records (EHRs), medical photographs, and research documents [4,5]. AI can automatically classify and label these assets, making finding and retrieving

DOI: 10.1201/9781032673479-8

them easier. This enhances efficiency for healthcare workers and mitigates the danger of errors linked to human data manipulation. AI-based Digital Asset Management (DAM) is crucial in enhancing diagnostic skills, data organization, and research and development in the healthcare industry. Pharmaceutical businesses and research institutions handle extensive datasets in drug discovery, clinical trials, and other research endeavours [6]. AI algorithms can analyse these statistics to discern patterns, connections, and potential insights that may elude human researchers [7]. This expedites the tempo of research, resulting in expedited discoveries and improvements in medical science. Using AI in digital asset management (DAM) is also significant in personalized medicine. AI algorithms can utilize patient data, such as genetic information, treatment histories, and lifestyle characteristics, to aid in the development of personalized treatment programmes for individual patients. Such a degree of personalization enhances the effectiveness of treatment and reduces the likelihood of adverse reactions, hence promoting patient-centred healthcare approaches [8,9]. Amidst the ongoing digital revolution of the healthcare business, ensuring the security and privacy of patient data remains of utmost importance. AI-powered digital asset management (DAM) solutions integrate strong security measures such as encryption, access limits, and anomaly detection to protect confidential medical data. This not only guarantees adherence to data protection requirements but also fosters confidence among patients regarding the privacy of their healthcare data [10]. Ultimately, AI-driven Digital Asset Management in the healthcare system signifies a fundamental change in how medical data is handled, examined, and utilized. Combining AI and DAM can improve diagnostic accuracy, simplify data organization, expedite research, and facilitate personalized therapy. This collaboration offers the prospect of a healthcare future that is more efficient, accurate and focused on the patients' needs. The ongoing advancement of technology is expected to profoundly affect healthcare, bringing about a period where data becomes a valuable asset in improving health outcomes for both people and groups [11].

6.2 MEANING OF DIGITAL ASSETS

Digital assets encompass all types of data or information stored in a digital format and have inherent worth. These assets can include various materials, such as text files, photos, audio files, videos, software, and other digital files. Within business and finance, digital assets encompass not just cryptocurrencies but also digital certificates and electronically recorded data [12]. The importance of digital assets stems from their function as essential resources for individuals, corporations, and organizations. In the era of digitalization, these resources are vital elements of communication, entertainment, and commerce. They may be readily duplicated, transported, and altered, offering unparalleled versatility and availability. Digital assets possess value beyond their content, encompassing intellectual property, brand reputation, and financial worth [13,14]. Digital asset management encompasses the systematic arrangement, storage, retrieval, and distribution of digital assets, ensuring efficient utilization and safeguarding. In today's world, where digital technologies play a crucial role in multiple industries, it is vital to comprehend and efficiently handle digital assets to use the opportunities presented by the digital realm entirely [15].

Irrespective of the field, such as creative industries, healthcare, finance, or others, it is crucial to acknowledge and enhance the utilization of digital resources to navigate the intricacies of the contemporary, technology-driven world effectively.

6.3 TYPES OF DIGITAL ASSETS

Digital assets refer to a diverse range of content and information that exist in digital format. These assets can be classified according to their distinct qualities and purposes. Below is a comprehensive summary of many categories of digital assets (Figure 6.1).

Digital assets refer to a diverse range of content and information that exist in digital format. These assets can be classified according to their distinct qualities and purposes. Below is a comprehensive summary of many categories of digital assets:

Textual Assets: These refer to many documents, such as Word files, PDFs, spreadsheets, and other items mainly composed of text.

E-books: Electronic books in several formats, such as EPUB or Kindle.

6.3.1 IMAGE ASSETS

Photographs: Digital photos acquired through cameras or generated by other devices.

Graphics: Creation of digital images, logos, icons, and various other visual components.

Infographics: Graphical illustrations that communicate information or data in a visual format.

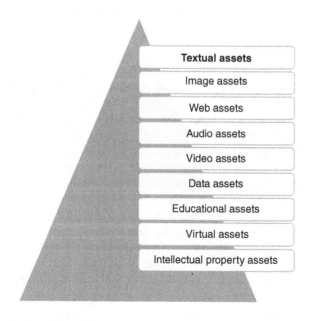

FIGURE 6.1 Types of Digital Assets.

6.3.2 Audio Assets

Music: Digital audio files stored in MP3, WAV, or FLAC formats.
Podcasts: Audio recordings that can be streamed or downloaded.
Sound Effects: It refers to digital files used to enhance multimedia material.

6.3.3 Video Assets

Videos: Dynamic visual content available in multiple forms, such as MP4, AVI, or MKV.
Animations: These refer to sequences of moving images typically generated using animation software.
Interactive Assets: It includes software programs and smartphone apps.
Games: It refers to digital interactive entertainment that frequently incorporates multimedia components.

6.3.4 Web Assets

Websites: Digital platforms that may be accessed through web browsers.
Social Media: This content refers to the various posts, photographs, and videos published on social media sites.
Cryptocurrencies, including Bitcoin and Ethereum, are digital or virtual currencies that rely on encryption for security.

6.3.5 Intellectual Property Assets

Trademarks: Digital depictions of trademark logos or symbols.
Protected Content: Digital files safeguarded by copyright rules.

6.3.6 Data Assets

Databases: Aggregations of organized digital information.
Big Data refers to extensive collections of digital information utilized to conduct analytics and gain valuable insights.

6.3.7 Virtual Assets

Virtual Real Estate: Digital areas within virtual surroundings.
Virtual Goods: Digital objects or things utilized within online games or virtual environments.
Social Media assets refer to profiles and digital representations of individuals or companies on social platforms.
Posts and Shares: Digital information that is distributed throughout social networks.

6.3.8 EDUCATIONAL ASSETS

Online Courses: Digital resources and instructional programmes for learning.
Educational Software: Digital solutions designed to facilitate learning and enhance skills development.

Efficiently overseeing these digital resources is vital for both corporations and individuals. Digital Asset Management (DAM) solutions are specifically developed to efficiently organize, store, and retrieve digital assets, ensuring effortless access and utilization across multiple digital platforms and channels [16].

6.4 SOFTWARE APPLICATIONS USED IN CREATING AND MANAGING DIGITAL ASSETS

Various domains utilize software applications and technologies to manage and create digital assets. Below are many crucial ones, along with concise explanations:

Digital Asset Management (DAM): It is a type of software that is used to manage and organize digital assets. DAM software is specifically developed to facilitate digital asset organization, storage, and retrieval. It offers a centralized storage for assets, enabling users to efficiently classify, search, and oversee information. Notable examples consist of Adobe Experience Manager, Widen Collective, and Bynder.

The Adobe Creative Cloud Suite: Adobe Creative Cloud encompasses software tools such as Photoshop, Illustrator, and InDesign. These apps produce and modify diverse digital assets, including photos, graphics, and documents.

Canva: Canva is an online graphic design platform that enables users to produce various types of visual content, such as social media graphics, posters, presentations, and more, without requiring advanced design expertise.

Final Cut Pro and Adobe Premiere Pro: These video editing software packages are commonly utilized for creating and modifying video footage. They offer sophisticated functionalities for video editing, such as trimming, merging, applying effects, and additional options.

Comparison between Audacity and Adobe Audition: Audacity is a software for editing freely available audio, whereas Adobe Audition is a high-quality tool for professional audio editing. Both tools are utilized to capture, modify, and improve audio resources.

Comparison of Unity with Unreal Engine: Unity and Unreal Engine are robust game development frameworks that create engaging and immersive digital assets, including video games, simulations, and virtual reality experiences.

Microsoft Office Suite: Microsoft Office Suite comprises apps such as Word, Excel, and PowerPoint, which are widely utilized for generating text, spreadsheets, and presenting digital resources.

Google Workspace: Google Workspace, previously called G Suite, is a collection of cloud-based productivity tools such as Google Docs, Sheets, and Slides. These tools allow users to simultaneously create, edit, and collaborate on digital files.

WordPress: WordPress is a widely used content management system for creating and administrating websites and blogs. It facilitates users in effortlessly publishing and organizing digital content.

Hootsuite and Buffer: These tools for social media management facilitate the scheduling, uploading, and analysis of social media content. They play a vital role in overseeing and enhancing social media resources on multiple platforms.

GitHub is a web-based platform allowing users to store and manage their code repositories. GitHub is a site designed for version control for managing and collaborating on software development projects. It facilitates the monitoring of modifications, the management of code versions, and the collaboration of digital assets within a team setting.

Each instrument fulfils a distinct function in managing digital assets, encompassing their creation, modification, organization, storage, and dissemination. The selection of tools is contingent upon the user or organization's particular requirements and operational processes [17,18].

6.5   CONTRIBUTORS OF AI IN DIGITAL ASSET MANAGEMENT

AI's progress in handling and generating digital assets has been a cooperative endeavour encompassing contributions from researchers, engineers, and organizations. Diverse individuals have contributed to the development of AI, a revolutionary force in Digital Asset Management (DAM). Machine learning is pivotal in facilitating the automated categorization, tagging, and organization of enormous quantities of digital assets by DAM systems [19]. This feature optimizes the asset management process and improves search functionalities, facilitating users in efficiently locating desired content. Natural Language Processing (NLP) is another noteworthy factor that enables DAM systems to comprehend and interpret textual data linked to digital assets, enhancing the content's accessibility and organization [20]. Additionally critical is the function of computer vision, which enables AI to analyse and identify visual components in images and videos, facilitating content recognition and enabling more sophisticated search capabilities [21]. Moreover, ongoing progress in deep learning facilitates the enhancement of AI algorithms, thereby augmenting their capacity to comprehend user preferences and behaviour. As a result, digital asset management solutions become more intuitive and personalized. These collaborative efforts advance the development of AI in DAM, resulting in more intelligent and user-friendly asset retrieval and organization [22,23]. Below is a list of noteworthy individuals who have made significant contributions, along with a description of their important achievements and the respective years in which they occurred (Figure 6.2).

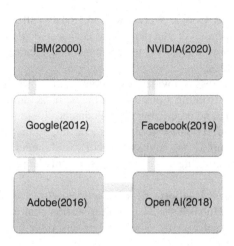

FIGURE 6.2   Organizations in AI based Digital Asset Management.

6.5.1   IBM (2000)

IBM has advanced AI, significantly contributing to developing AI technologies, specifically in data management and analytics. During the 2000s, IBM's Watson, a cognitive computing system, achieved a notable milestone by demonstrating the promise of AI in NLP and data analysis.

6.5.2   Google (2012)

In 2012, Google achieved a significant advancement by introducing the deep learning algorithm of Google Brain. The Google Net algorithm has shown exceptional abilities in picture identification, significantly impacting the advancement of AI applications for visual data, such as digital images and videos.

6.5.3   Adobe (2016)

Adobe has made a significant contribution to the integration of AI into the process of creating digital assets. Adobe introduced Adobe Sensei, an AI and machine learning platform, 2016 to improve its Creative Cloud products. Sensei plays a role in generating content, automating processes, and providing tailored experiences in digital resources.

6.5.4   OpenAI (2018)

OpenAI has played a significant role in enhancing AI capabilities. In 2018, OpenAI unveiled GPT-2 (Generative Pre-trained Transformer 2), a language model significantly enhancing natural language comprehension. This has influenced the progress of AI applications for digital assets based on text.

6.5.5 FACEBOOK (2019)

Facebook significantly advanced AI in digital assets by developing the PyTorch platform. Poarch has been popular because of its versatility and user-friendly nature, which simplifies the creation of AI applications, particularly those involving image and video processing.

6.5.6 NVIDIA (2020)

NVIDIA has played a leading role in the advancement of AI hardware development. During the 2020s, GPUs (graphics, etc.) have become widespread for speeding up AI tasks, particularly in image and video processing in digital asset management. These contributions only reflect a small portion of the joint endeavours shaping the convergence of AI and digital assets. The advancement of AI remains a collaborative effort involving researchers, institutions, and industry leaders pushing the limits of what can be achieved in digital asset management and development [24].

6.6 IMPORTANCE OF AI IN DIGITAL ASSET MANAGEMENT

AI is significant in digital asset management (DAM) since it fundamentally transforms how organizations manage, analyse, and extract value from their digital assets. AI is essential in the context of DAM for the following reasons:

6.6.1 OPTIMAL ORGANIZATION AND RETRIEVAL

AI improves digital asset management (DAM) systems by autonomously labelling, classifying, and arranging enormous quantities of digital resources. This optimizes the asset recovery, rendering it more efficient and precise than conventional manual techniques. The effective management and retrieval of digital assets are fundamental components of Digital Asset Management (DAM). In this regard, incorporating AI is instrumental in augmenting these operations [25,26]. AI technologies enhance the organization of extensive repositories of digital assets with increased efficiency and accuracy. DAM systems can autonomously execute machine learning algorithms to assess, classify, and annotate digital assets by content, context, and user engagement. This approach guarantees a more uniform and precise resource arrangement by minimizing the manual labour needed for metadata labelling [27].

6.6.2 AUTOMATED METADATA GENERATION

AI systems can autonomously provide metadata for digital objects, encompassing keywords, descriptions, and contextual information. This expedites the process of categorizing and guarantees uniform and standardized information, enhancing the precision of searches.

6.6.3   Improved Search and Exploration

AI in digital asset management (DAM) systems enhances the search functionality, allowing users to locate pertinent assets more efficiently. NLP and picture recognition technologies enhance the search experience by making it more intuitive and precise, resulting in time savings and improved productivity.

6.6.4   Customized Tailoring

AI algorithms possess the capability to scrutinize user behaviour and preferences in order to provide tailored content suggestions. This is especially advantageous in marketing and customer engagement, as customized material improves user experience and stimulates active participation.

6.6.5   Automated Workflows

AI streamlines repetitive activities and workflows in DAM systems, including file conversions, format standardization, and distribution. This process automation diminishes the need for human labour, mitigates mistakes, and expedites the total progression.

6.6.6   Enabling and Guiding the Development of Innovative Processes

AI capabilities incorporated into Digital Asset Management (DAM) systems aid in the execution of artistic tasks, such as the manipulation of images and videos. Features such as automated image enhancement, object detection, and style transfer boost the efficiency and creativity of content creation.

6.6.7   Advanced Data Analysis and Understanding

AI facilitates advanced analysis of digital resources, offering valuable information on usage patterns, performance metrics, and user interaction. These analytics enable organizations to make data-based decisions and optimize their content strategy.

6.6.8   Adapting to Changing Trends Flexibly and Responsively

AI algorithms can detect and analyse trends and patterns in the use of digital resources, enabling organizations to adjust their content strategy promptly. Adaptability is vital in rapidly changing businesses where maintaining relevance is critical.

6.6.9   Ensuring Safety and Adherence to Regulations

AI safeguards digital assets by implementing access controls, encryption, and anomaly detection. Ensuring compliance with data protection standards and safeguarding sensitive information is paramount. Beyond enhancing efficiency and organization,

AI in Digital Asset Management (DAM) is of the utmost importance in guaranteeing safety and regulatory compliance [28,29]. Compliance is substantially aided by AI technologies, which automate and bolster security measures within DAM systems. Algorithms capable of machine learning can identify and alleviate potential security risks, including suspicious activities and unauthorized access, thereby protecting digital assets from unauthorized manipulation or use.

Moreover, AI in DAM facilitates regulatory compliance by automating the application of metadata identifiers, content categorization, and access controls. This practice guarantees the proper management of sensitive or regulated content, thereby mitigating the potential for non-compliance with data protection laws or industry-specific regulations [30,31]. DAM systems can identify and manage potentially sensitive information by leveraging AI's capability to analyse and interpret textual and visual content; this feature strengthens the safeguarding of intellectual property and ensures adherence to copyright regulations [32].

6.6.10 Scalability and Managing Large Volumes of Data

Digital Asset Management (DAM) systems frequently handle vast quantities of data. AI allows these systems to expand and process large amounts of data, guaranteeing that organizations can effectively manage and analyse their digital assets regardless of volume. Incorporating AI into Digital Asset Management systems fundamentally changes how organizations handle digital content. It allows them to manage their digital assets and gain valuable insights efficiently, enabling them to foster innovation in a world increasingly centred around digital technologies [33]. Scalability and the capacity to effectively handle substantial quantities of data are critical factors that underscore the significance of AI in Digital Asset Management (DAM) systems. Conventional approaches to classifying and organizing them by hand have become impracticable in light of the exponential growth of digital asset volume. AI overcomes this obstacle by providing scalable solutions that automate the processing and administration of enormous quantities of data. AI algorithms, specifically those powered by machine learning, enable DAM systems to manage a wide range of datasets effectively [34,35]. Digital assets can be automatically analysed, categorized, and tagged using these algorithms, which rely on user interactions, content characteristics, and patterns. This process not only expedites the arrangement of resources but also guarantees uniformity and precision, thereby reducing the likelihood of human fallibility linked to manual administration.

6.7 LIMITATIONS

The possibility of bias in AI algorithms, which could lead to skewed categorizations and have an effect on the diversity and inclusiveness of content representation, is an additional concern. Maintaining equilibrium between the advantages offered by AI-powered automation and tackling these constraints continues to be a pivotal element in the responsible utilization of AI in the realm of digital asset management [36,37]. Some of the limitations are as follows (Figure 6.3).

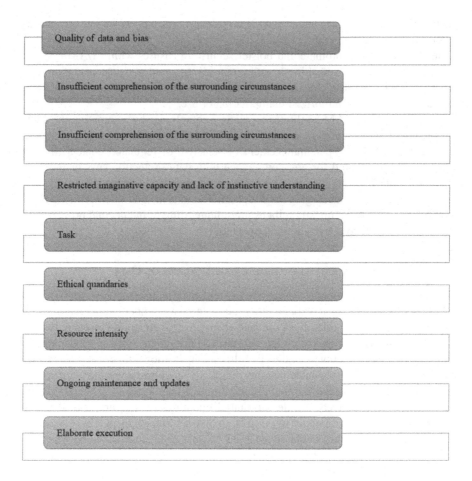

FIGURE 6.3   Limitations of AI based Digital Asset Management.

6.7.1   QUALITY OF DATA AND BIAS

The effectiveness of AI algorithms is greatly influenced by the calibre and variety of the data used for training. Biased or incomplete training data for AI models can lead to erroneous or skewed findings, which can negatively affect the effectiveness of digital asset management.

6.7.2   INSUFFICIENT COMPREHENSION OF THE SURROUNDING CIRCUMSTANCES

AI systems may encounter difficulties comprehending the context and subtleties of certain digital assets. This constraint might result in misunderstandings and inaccuracies in tasks that require a profound comprehension of specific sectors, cultural allusions, or intricate topics.

6.7.3 Insufficient Comprehension of the Surrounding Circumstances

Numerous organizations employ a diverse range of digital asset management methods and systems [38]. Attaining smooth compatibility among these varied systems, particularly when including AI, might provide a substantial obstacle. Compatibility difficulties may occur, impeding the seamless transfer of data and resources.

6.7.4 Restricted Imaginative Capacity and Lack of Instinctive Understanding

Although AI demonstrates proficiency in repetitive and data-driven activities, it frequently needs to gain humans' creative and intuitive abilities in content creation and decision-making. This constraint may be particularly evident in sectors that largely depend on artistic or subjective components.

6.7.5 Task

Incorporating AI into digital asset management presents novel security obstacles. AI models are vulnerable to adversarial attacks, in which malevolent individuals modify the input data to deceive the system [39]. Safeguarding the security of valuable digital assets is an essential factor to consider.

6.7.6 Ethical Quandaries

AI systems have the potential to unintentionally maintain or intensify preexisting biases that are already present in the training data. This raises ethical considerations, particularly in applications like content recommendation, where biased algorithms can perpetuate stereotypes or discriminatory practices.

6.7.7 Resource Intensity

The process of training advanced AI models requires a significant amount of processing power and resources [40]. Smaller organizations or those with insufficient computational resources may need help utilizing advanced AI capabilities in digital asset management, which can impede their capacity to benefit from such skills.

6.7.8 Ongoing Maintenance and Updates

AI models necessitate periodic updates to accommodate shifting data trends and technology advancements. Organizations face resource and time problems in maintaining AI algorithms' relevance and accuracy, necessitating continuous effort.

6.7.9 Elaborate Execution

Incorporating AI into preexisting digital asset management systems can be intricate. The adoption process may include substantial modifications to processes, personnel training, and infrastructure, which could result in opposition and significant interruptions [41]. Comprehending and resolving these constraints is essential for organizations seeking to utilize AI in digital asset management to their advantage correctly. Ensuring a harmonious integration of AI capabilities with the necessary human skills is crucial for fully harnessing the potential of these technologies.

6.8 CONCLUSION

Lastly, incorporating AI with digital asset management (DAM) signifies a revolutionary period in healthcare and other industries [42,43]. The synergy between AI and digital asset management (DAM) maximizes healthcare protocol effectiveness, improves diagnoses' precision, and streamlines tailored patient treatment. AI's capacity to analyse extensive quantities of medical data, namely in medical imaging, accelerates the decision-making process for healthcare practitioners, leading to enhanced patient outcomes [44,45]. In addition to healthcare, AI-powered Digital Asset Management (DAM) plays a crucial role in organizing data, conducting research, and facilitating development, leading to faster breakthroughs in pharmaceuticals and medical science. The widespread use of digital assets in several industries has made proper administration of these assets essential [46]. AI-supported Digital Asset Management (DAM) solutions optimize organizing, retrieving, and securing assets while assuring compliance with data protection regulations. The importance of AI in DAM is emphasized by its capacity to provide efficient asset categorization, streamline workflows, and conduct sophisticated data analysis, enabling organizations to respond to evolving trends and assure scalability effectively [47,48]. Nevertheless, the integration of AI in DAM encounters certain obstacles. Constraints arise from biased data, interoperability difficulties, and ethical considerations [49]. Security issues, resource intensity, and continuing maintenance requirements further complicate the integration process. Organizations need to tackle these problems to fully exploit AI's potential advantages in DAM and successfully navigate the changing terrain of digital asset management. In the end, the combination of AI and DAM has the potential to create a future in healthcare and other fields that is more efficient, inventive, and focused on the needs of patients [50].

REFERENCES

1. Miller D.D., Brown E.W. Artificial intelligence in medical practice: the question to the answer? *Am J Med.* 2018;131(2):129–133.
2. Kirch D.G., Petelle K. Addressing the physician shortage: the peril of ignoring demography. *JAMA.* 2017;317(19):1947–1948.
3. Combi C., Pozzani G., Pozzi G. Telemedicine for developing countries. *Appl Clin Inform.* 2016;7(4):1025–1050.
4. Bresnick J. *Artificial intelligence in healthcare market to see 40% CAGR surge,* Xtelligent Healthcare Media, Massachusetts, United States; 2017.
5. Lee K.-F. *AI superpowers: China, Silicon Valley, and the new world order,* 1st ed. Houghton Mifflin Harcourt, Boston, Massachusetts ; 2019.
6. King D. D eepMind's health team joins Google Health. https://blog.google/technology/health/deepmind-health-joins-google-health/

7. Hoyt R.E., Snider D., Thompson C., Mantravadi S. IBM Watson Analytics: automating visualization, descriptive, and predictive statistics. *JMIR Public Health Surveill.* 2016;2(2):e157.

8. Marr B. *How is AI used in healthcare-5 powerful real-world examples that show the latest advances.* Forbes, New Jersey, US; 2018.

9. Kalis B., Collier M., Fu R. *10 promising AI applications in health care.* Harvard Business Review; Brighton, Massachusetts 2018.

10. Singhal S., Carlton S. *The era of exponential improvement in healthcare?* McKinsey Co Rev.; New York, USA 2019.

11. Konieczny L., Roterman I. Personalized precision medicine. *Bio-Algorithms Med-Syst.* 2019;15, 513–5537.

12. Love-Koh J. The future of precision medicine: potential impacts for health technology assessment. *Pharmacoeconomics.* 2018;36(12):1439–1451.

13. Kulski J.K. *Next-generation sequencing-an overview of the history, tools, and 'omic' applications*; Intechopen, London, UK 2020.

14. Hughes J.P., Rees S., Kalindjian S.B., Philpott K.L. Principles of early drug discovery. *Br J Pharmacol.* 2011;162(6):1239–1249.

15. Ekins S. Exploiting machine learning for end-to-end drug discovery and development. *Nat Mater.* 2019;18(5):435–441.

16. Zhang L., Tan J., Han D., Zhu H. From machine learning to deep learning: progress in machine intelligence for rational drug discovery. *Drug Discov Today.* 2017;22(11):1680–1685.

17. Lavecchia A. Deep learning in drug discovery: opportunities, challenges and future prospects. *Drug Discov Today.* 2019;24(10):2017–2032.

18. Coley C.W., Barzilay R., Green W.H., Jaakkola T.S., Jensen K.F. Convolutional embedding of attributed molecular graphs for physical property prediction. *J Chem Inf Model.* 2017;57(8):1757–1772.

19. Mayr A., Klambauer G., Unterthiner T., Hochreiter S. DeepTox: toxicity prediction using deep learning. *Front Environ Sci.* 2016;3:80.

20. Wu Z. Molecule Net: a benchmark for molecular machine learning. *Chem Sci.* 2018;9, 513–530.

21. Kadurin A., Nikolenko S., Khrabrov K., Aliper A., Zhavoronkov A. druGAN: an advanced generative adversarial autoencoder model for de novo generation of new molecules with desired molecular properties in silico. *Mol Pharm.* 2017;14(9):3098–3104.

22. Blaschke T., Olivecrona M., Engkvist O., Bajorath J., Chen H. Application of generative autoencoder in de novo molecular design. *Mol Inform.* 2018;37(1–2):1700123.

23. Merk D., Friedrich L., Grisoni F., Schneider G. De novo design of bioactive small molecules by artificial intelligence. *Mol Inform.* 2018;37, 1700153.

24. Shi T. Molecular image-based convolutional neural network for the prediction of ADMET properties. *Chemom Intell Lab Syst.* 2019;194:103853.

25. Wallach H.A., Dzamba M.I. AtomNet: a deep convolutional neural network for bioactivity prediction in structure-based drug discovery. *arXiv.* 10, 1–11 2015.

26. Hashimoto D.A., Rosman G., Rus D., Meireles O.R. Artificial intelligence in surgery: promises and perils. *Ann Surg.* 2018;268:70–76.

27. Petscharnig S., Schöffmann K. Learning laparoscopic video shot classification for gynecological surgery. *Multimed Tools Appl.* 2018;77:8061–8079.

28. Lundervold A.S., Lundervold A. An overview of deep learning in medical imaging focusing on MRI. *Z Med Phys.* 2019;29:102–127.

29. Mauro A.D., Greco M., Grimaldi M. A formal definition of big data based on its essential features. *Libr Rev.* 2016;65(3):122–135.

30. Doyle-Lindrud S. The evolution of the electronic health record. *Clin J Oncol Nurs.* 2015;19(2):153–154.

31. Reisman M. EHRs: the challenge of making electronic data usable and interoperable. *Pharm Ther.* 2017;42(9):572–575.
32. Murphy G., Hanken M.A., Waters K. *Electronic health records: changing the vision.* Philadelphia, PA: Saunders W B Co; 1999, p. 627.
33. Moore S.K. Unhooking medicine [wireless networking]. *IEEE Spectr.* 2001;38(1):107–108, 110.
34. Laney D. *3D data management: controlling data volume, velocity, and variety, application delivery strategies.* Stamford, CA: META Group Inc; 2001.
35. Gillum R.F. From papyrus to the electronic tablet: a brief history of the clinical medical record with lessons for the digital Age. *Am J Med.* 2013;126(10):853–857.
36. Nasi G., Cucciniello M., Guerrazzi C. The role of mobile technologies in health care processes: the case of cancer supportive care. *J Med Internet Res.* 2015;17(2):e26.
37. Gulshan V., Peng L., Coram M., Stumpe M.C., Wu D., Narayanaswamy A., et al. Development and validation of a deep learning algorithm for detection of diabetic retinopathy in retinal fundus photographs. *JAMA.* 2016;316(22):2402.
38. Esteva A., Kuprel B., Novoa R.A., Ko J., Swetter S.M., Blau H.M., Thrun S. Dermatologist-level classification of skin cancer with deep neural networks. *Nature.* 2017;542(7639):115–118.
39. Lazer D., Kennedy R., King G., Vespignani A. The parable of Google Flu: traps in big data analysis. *Science.* 2014;343:1203.
40. Davis K., Stremikis K., Schoen C., Squires D. *Mirror, mirror on the wall, 2014 update: how the U.S. Health Care System compares internationally.* The Commonwealth Fund; New York, USA 2014.
41. JASON. *Perspectives on research in artificial intelligence and artificial general intelligence relevant to DoD*; Mitre Corporation, Mclean, Virginia 2017.
42. Dias D., Cunha J.P.S. Wearable health devices-vital sign monitoring, systems and technologies. *Sensors (Basel).* 2018;18(8):2414.
43. Becker A. Artificial intelligence in medicine: what is it doing for us today? *Health Policy Technol.* 2019;9:198–205.
44. Boris Wertz, 2016, Data, not algorithms, is key to machine learning success. https://medium.com/machine-intelligence-report/data-not-algorithms-is-key-to-machine-learningsuccess-69c6c4b79f33.
45. Patnab, 2016, 10 Great Healthcare Data Sets. https://www.datasciencecentral.com/profiles/blogs/10-great-healthcare-data-sets.
46. https://www.image-net.org/challenges/LSVRC/2017/index.php.
47. Gideon Lewis-Kraus, 2016, The great A.I. awakening. https://www.nytimes.com/2016/12/14/magazine/the-great-ai-awakening.html.
48. https://www.cbinsights.com/research/artificial-intelligence-startups-healthcare.
49. Dillon Baker, 2017, Understanding the artificial intelligence hype cycle. https://contently.com/strategist/2017/05/23/artificial-intelligence-hype-cycle-5-stat.
50. John Mannes, 2016, PlushCare nabs $8M series A to prove telehealth can go mainstream. https://www.directderm.com/, and PlushCare: https://techcrunch.com/2016/11/03/plushcare-nabs-8m-series-a-to-prove-telehealth-can-go-mainstream/.

7 Role of IoT and Big Data in Evolving Smartness in Manufacturing and Lifestyle

Nidhi Chahal, Arun Kumar Rana,
Preeti Bansal, Simarpreet Kaur, Himanshi Sachan,
Shresth Modi, and Amit Kumar Kesarwani

7.1 INTRODUCTION

The Internet of Things (IoT) and Big Data are now the most debated and discussed technology, when it comes to data analysis. IoT devices, which are the major source of collecting data through various sensors and in-built components attached to the device, transfer the valuable data through Internet to the cloud. These valuables are categorized as Big Data which are then compiled and analyzed by Artificial Intelligence and Machine Learning to generate useful and insightful information for the Business and Organization. The IoT are the devices that give us the feasibility and compatibility to transfer data over a network of interconnected devices or a connection between the user and the devices with the least human intervention.

These data-collecting devices or so-called 'things' are categorized into three parts:

- **Things That Collect Information and Send It to Cloud Over Internet**: IoT devices are integrated devices in which various sensors like Temperature sensors, Air Quality sensor, Soil Moisture sensors, and many other sensors are embedded together to get the data from the environment to process it into Information.
- **Things That Receive Data and Act on It**: Machinery like PCB Manufacturing, Semi-conductor chip manufacturing machines, or even a daily use printer machine get the data from the various sensors and equipment for analyzing that data in order to perform the desired task.
- **Things That Can Perform Both These Tasks**: Both the above "Things" can be brought together to perform both task in one go [2].

These tremendous sums of information collected by the gadgets or so-called "Things" are complexed and can be troublesome to prepare with conventional strategy.

DOI: 10.1201/9781032673479-9

These information are both organized and unstructured information, known as Enormous Data [1]. These complex information sets shape modern information sources that are so voluminous that conventional information-preparing computer programs cannot oversee them, but this gigantic volume of information can be utilized to address the Commerce Issue with unused inventive thoughts. The Enormous Information works on three Vs. These five Vs are Volume, Velocity, Variety, Value, and Veracity, alluding to Figure 7.1 [3].

- **Volume:** Information volume is noteworthy. Preparing colossal sums of low-density, unstructured data could be a when working with huge information. These can be unvalued information sources like Twitter information nourishes, clickstreams from websites or portable apps, or sensor-enabled equipment. It may be tens of gigabytes of information for certain enterprises. A few hundred petabytes might apply to others.
- **Velocity:** Speed is the fast rate at which data is obtained and (maybe) utilized. The most noteworthy speed of information regularly streams into memory without being replicated to disk. A few Internet-enabled shrewd contraptions work in genuine time or exceptionally near to it, requiring real-time examination and response.
- **Variety:** Assortment implies to the wide run of information sorts that are open. In a social database, conventional information sorts were organized and effectively suited. Information presently arrives in unused unstructured information designs that are much appreciated for the development of huge information. Content, sound, and video are illustrations of semi-structured

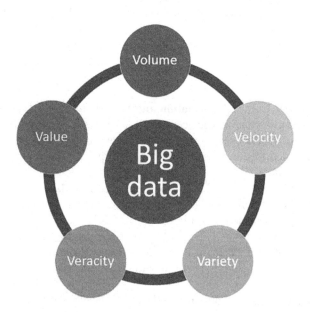

FIGURE 7.1   The five Vs of Big Data.

and unstructured information groups that require assistance preprocessing in order to supply meaning and empower metadata [4].

- **Value:** In spite of the convenience inalienable in huge information, it must be determined. Capable expository apparatuses, learned information researchers, and administrators who are responsive to the data's message are essential for recognizing the esteem in an organization's information collection. One field that has the potential to be exceptionally important is prescient analytics, which helps officials estimate results based on designs they spot within the data.
- **Veracity:** Although big information can be deluding due to its complexity, actualities don't lie. Information researchers must pay attention to evacuate any botches, copy passages, and other variations from the norm that can meddle with their capacity to reveal important insights.

Market leaders all around the world use the interdependence between IoT and Big Data to their advantage, benefiting from innovation and providing them a competitive edge. We now have more gadgets linked to the Internet and to one another than ever before because connection is becoming more and more important in our daily working and social lives. To store, acquire, and manage data from the continuously streaming data from the IoT-enabled devices, the Big Data technologies need to be enhanced. This illustrates the interlinked results of IoT and Big Data together, creating a world of deep knowledge and insightful information processed through various unique Artificial Intelligence and Machine Learning algorithms [2]. In this perception, Big Data is used to power the IoT. These two ideas will be used to help build a contemporary, intelligently linked world. IoT's involvement will highlight the advantages of making wiser decisions, making the digital world both smarter and safer.

7.1.1 How Do IoT and Big Data Impact Each Other?

IoT and Big Data are interdependent and have a significant impact on one another. As IoT develops, there is a corresponding rise in demand for Big Data capabilities, and as data volumes rise daily, organizations need to update their Big Data storage infrastructure. Due to this interdependence of IoT and Big Data, as IoT expands quickly, traditional data storage is under increasing strain, which encourages the development of cutting-edge Big Data solutions. In order to meet this expanding demand, businesses are compelled to improve their technology and systems. By transforming data into something useful for businesses, Big Data and the IoT ultimately share the same objectives and depend on one another to accomplish them. The insights and knowledge gleaned from these technologies will enhance the analytics process, making it quicker and simpler, improving efficiency, saving money, and encouraging well-informed forecasts [7].

7.1.2 Interaction of Internet of Things (IoT) and Big Data

The interaction between the IoT and Big Data is a powerful synergy that has the potential to revolutionize the way we collect, analyze, and utilize data. Figure 7.2

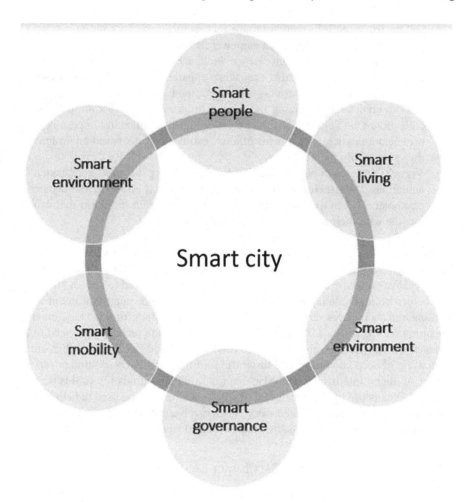

FIGURE 7.2 Symbiotic interaction of IoT and Big Data.

shows the symbiotic relationship between IoT and Big Data. The application of this synergic interaction of technologies is seen in various manufacturing and smart industries. Both technologies intersect and complement each other in different fields and functionalities.

1. **Data Generation and Collection:** Smart and sensor-equipped IoT devices produce enormous volumes of data in real time. These gadgets gather information on temperature, humidity, location, and a variety of other factors in a variety of locations, including homes, workplaces, healthcare facilities, and agricultural settings. Big Data Technology is used to handle this vast number of data, both structured and unstructured. They offer the facilities required to handle and store the enormous datasets produced by IoT devices.

2. **Data Storage:** Big Data solutions, such as distributed file systems (e.g., Hadoop Distributed File System), NoSQL databases, and data warehouses, are well-suited for handling the scalability and storage requirements of IoT-generated, continuous stream of data.

3. **Data Processing and Analytics:** Big Data analytics tools, like Apache Spark and Apache Flink, provide the capability to process and analyze large datasets in real time or batch mode, which is crucial for IoT applications as these analyzed data are further used for extracting meaningful insights.

4. **Data Integration:** The data generated by the IoT devices are of various types: Structured as well as Unstructured. Big Data is designed in such a way to analyze and process those data.

5. **Machine Learning and Predictive Analytics:** Machine Learning is changing the way data are analyzed to predict future forecast events or behaviors based on historical data. Such a platform of computational power and storage required for training and deploying is provided by Big Data.

6. **Security and Privacy:** Given the high number of endpoints associated with IoT devices, security is a major problem. Because some IoT data is sensitive, privacy considerations also exist. Big Data systems frequently include security safeguards like encryption and access limits. They are essential in handling and protecting the enormous volumes of data that IoT devices create.

In conclusion, the relationship between IoT and Big Data is mutually beneficial. Big Data offers the infrastructure and technologies required to manage, analyze, and extract useful insights from the vast volumes that IoT creates. Realizing the full potential of these technologies to build a more connected, intelligent, and effective society depends on their cooperation [9]. Hence, based on data analysis and IoT infrastructures, "Smart Cities" as a general application that includes smart grids, smart transportation, smart manufacturing, smart buildings, and much more, have become more popular.

The remaining part of the chapter is organized as follows:

- **Topic 2: History:** This topic relics the origin and evolution of IoT and Big Data from the late 18th century to its revolution and technological transformation in the 21st century.
- **Topic 3: Methodology:** This chapter is structured into the discussion of the impact of IoT and Big Data in various fields and imprints the advantages and challenges of the integration of such technologies.
- **Topic 4: SMART TRANSPORTATION: Navigating Tomorrow's Roads** discusses the role of the Synergic integration of IoT and Big Data in transforming the transportation infrastructure.
- **Topic 5: SMART HEALTHCARE: A New Way of Diagnosis** presents a new way of patient's diagnosis using the advancement of technologies and analysis of data from smart electronic devices.
- **Topic 6: SMART GRID: Sustainable Energy, Efficient consumption** presents the sustainable utilization of energy resources for efficient power generation, consumption, and its distribution.

- **Topic 7: SMART INVENTORY: Changing the Warehouse Mechanism** discusses the integration of technologies such as RFID, IoT sensors, and analytics to streamline and automate inventory management, aiming to enhance accuracy, reduce costs, minimize stockouts, and optimize supply chains for businesses through intelligent data-driven decision-making.
- **Topic 8: SMART CITIES: A New Lifestyle of Living with Easiness** deliberates the ever-changing lifestyle and demand of smart facilities integrated with IoT devices in order to ease the living of the urban population in smart cities.
- **Topic 9: Conclusion and Future Scope:** The contribution of this chapter, along with some important conclusions, outlines of future research directions, have been summarized.

7.2 HISTORY

Machine have been giving coordinated communication since the transmission was presented in the 1830s and 1840s. Portrayed as "Wireless telegraphy," the primary radio wave transmission was put on 3 June 1900, giving an essential component for creating the IoT. But the following of beginnings of the Web of Things go to 1980s, when a few college understudies chose to adjust a Coca-Cola distributing machine to empower inaccessible observing of its contents. This was the primary time that the concept of implanting sensors and insights into physical objects was put out. But the pace of specialized advancement was drowsy and difficult. In spite of the fact that it was as it were when Kevin Ashton, a computer researcher, to begin with, utilized the term "Web of Things" in 1999. While working at Procter and Bet, Ashton advanced the utilize of radio-frequency recognizable proof (RFID) chips to track things as they move through the supply chain. Since that point, IoT innovation has progressed essentially and helping in different domains., . Over the course of the resulting decade, the number of associated gadgets on the showcase expanded, starting a surge in the open intrigue in IoT technology. The primary iPhone was presented in 2007, the primary savvy fridge was displayed by LG in 2000, and by 2008, there were more associated contraptions than there were individuals on the planet [5]. With the progression within the field of Web of Things innovation, the information created by IoT gadgets was becoming complex and tremendous. The information created by these IoT gadgets was of diverse frames, that is, organized, semi-structured, and unstructured, and such a huge information collection from which data is extricated for the advantage of organization and commerce is Enormous Information. In spite of the fact that the following information investigation that driven today's progressed huge information analytics begins way back in the 17th century in London, its integration with the Internet age as it was started within the late 1990s. It was in 1989 when Tim Berners-Lee and Robert Cailliau found the World Wide Web and created HTML, URLs, and HTTP while working for CERN. Through such unused development within the field of the web, the way of putting away data altered viably in late 1996 from paper to computerized. Sometime recently, during the look engine's presentation in 1997, the space google.com was registered [7]. This checked the starting of the look engine's climb to supremacy and the creation of endless extra specialized progressions, counting those within the areas of machine learning, huge information, and analytics. At that point and presently within the 21st century, enormous information detonated through the resourcefulness that it brings with it. In 2001, Doug Laney, an investigator at firm Gartner, gave the 3V – volume, variety , and velocity, characterizing the measurements and properties of enormous information. The Vs usher in a modern time in which enormous information may be considered a characterizing characteristic of the 21st century and capture the real substance of the term. Since that point, these Vs have been included in the list, counting honesty, esteem, and changeability. With the increment within the information utilization and everything getting robotized and a parcel of information produced and compiled, AWS (Amazon web administrations) in 2006 began advertising computing foundation administrations, now known as cloud computing. As of now, AWS is overwhelming the cloud administration industry with generally one-third of the worldwide advertise share. Worldwide economy and trade started going out to other cities and nations or getting to be universal, expanding their reach to the consumer of diverse locale and ethnicity, hence expanding the information set created and need of analyzing that for the improvement of organization benefits. When McKinsey anticipated that there would be a talent scarcity for analytics within the Joined together States by 2018, it was as of now clear that the industry would be in frantic require of such investigators require between 140,000 and 190,000 people with solid expository capacities and an advance 1.5 million investigators and supervisors who are competent of making precise data-driven judgments [11]. Facebook, moreover, set up the Open Compute Venture to trade specs for data centers that are vitality efficient. The exertion points to supply a 24% decreased fetched whereas expanding vitality proficiency by 38%. With the private organizations, the government organization too began realizing the requirements of the time. The Huge Information Investigate and Advancement Activity is propelled by the Obama administration with a $200 million commitment, citing the have to be the capacity to extricate beneficial experiences from information, speed up STEM (science, innovation, building, and mathematics) development, move forward national security, and change instruction. Since that point, the shortened form has changed to STEAM by grasping the expressions and including an A [8]. According to *Harvard Business Review*, the data scientist has become the foremost job in the century. Demand for data scientists increased as more businesses realized they needed to sift and extract information from unstructured data. For the primary in 2013, the global market was booming and reached 10 billion dollars alone for Big Data. With the reach of smartphone devices to everyone's hand, IBM reported that 2.5 quintillion (1,018) bytes of data is created every day. According to Allied Market Research, the Big Data and business analytics market reached $193.14 billion in 2019 and is projected to reach $420.98 billion by 2027 at a compound yearly growth rate of 10.9% [8]. For important economic sectors, edge computing is expected to change how data is handled and processed. Long-term Big Data lies in edge computing, which is defined as computing that is carried out close to the location where the data was collected as opposed to within the or a centralized data center.

7.3 METHODOLOGY

The methodology used in this article is based on an analytical review of available sources in a wide base of data such as conference papers, studies published in SSCI, IEEE, and Open Access journals, industrial reports, and reviews regarding the current state of the art of IoT and Big Data and its impact and aftermath on the various fields like Smart Manufacturing Industries, Smart Transportation, Smart Healthcare Sector, Smart Power Grid, Smart Inventory System and Smart Cities and the footprint of these synergic technologies on the economy and society. The research also discusses the challenges of integrating IoT applications with the industries.

- It explores the transforming change in the Transportation sector, making smart transportation.
- In this, they discuss how the interaction between IoT and Big Data improves the patient's diagnosis procedure.
- It envisions the scope of sustainable resource use by considering a smart grid mechanism.
- It finds a new and innovative way for industries to manage their Inventory system for more feasibility and less human exploitation.
- It visualizes the look-up of the development of urban cities into smart cities with integrated facilities for a better lifestyle.

7.4 SMART TRANSPORTATION-NAVIGATING TOMORROW'S ROADS

Global population is increasing abruptly thus increasing urbanization with the inflow of people into the cities. With the increase in population there is an increase in the usage of smart devices, cars, motorbikes, etc., which also raises many issues including pollution, traffic congestion, and resource management. [27]. But with the development of IoT, there are a massive number of IoT devices that are connected to the network. These gadgets continually gather data and send it to computer nodes for additional examination. Numerous applications use deep learning to evaluate the gathered data and gain "intelligence" and "automation" as a result of the substantial advancement of deep learning techniques. The dependency of people on transportation is increasing in daily life, increasing the traffic, and according to the estimates, traffic congestion costs a fortune for the economy. Just for a day, Bengaluru's economy alone is suffering a huge loss of 20,000 crore annually due to traffic congestion despite having 60 fully functional flyovers. The center of Bengaluru's economy, the IT segment, has been seriously affected by this occasion. Workers lose a part of their efficiency because of the length of time they are kept in activity. Agreeing to the overview, traffic-related challenges have fetched the IT division alone over Rs 7,000 crore. Moreover, the consideration emphasizes the negative results of activity blockage on small and medium-sized firms (SMEs) [18]. These companies battle to meet conveyance due dates, coming about in late shipments and troubled customers. To address these challenges, the think about prescribes a multi-pronged approach. This incorporates investigating underground transportation frameworks, especially for metros and government buses, and disheartening roadside stopping to prioritize street space for activity and guarantee person-on-foot security. The report also highlights the requirement for more noteworthy open travel, such as high-capacity buses, monorails, and metros, while debilitating the utilization of private transportation frameworks. The report, moreover, highlights the need for more prominent open travel, such as high-capacity buses, monorails, and metros, disheartening the utilization of private transportation frameworks. In this manner, part of IoT increments is to screen, analyze, and get real-time information. Unlike for open transportation, businesses are moreover seeking out solutions for transportation that can offer assistance to them to extend supply chain visibility, move forward operations at each arrange of coordination, and spare assets. This may be accomplished by collecting information about a coordination handle utilizing IoT gadgets and changing them into profitable trade data. Let's take a look at how the IoT in the transportation sector generates a constant supply of beneficial data so that businesses may make the most of Big Data [13].

7.4.1   HOW IoT CREATES VALUE FROM DATA?

It takes more than basically organizing hardware or sensors to actualize the Web of Things in transportation. The part of computerization utilizing IoT in Activity Administration can be seen in Figure 7.3. Each thing within the supply chain becomes a source of data that is much obliged to IoT in transportation. You would like be able to translate enormous information in coordination and supply chain administration in arrange to apply the gathered data for the advantage of a firm. Data that's not organized has small esteem. To develop your firm, you must be able to bargain with data [23]. In the event that information takes after the way of collect-communicate-aggregate-analyze-act, they ended up significant. IoT sensors accumulate information and transfer it to the cloud, where it is organized and put into a valuable way. The establishment for preparing ML calculations that assess information and robotize a few exercises is "filtered" data.

7.4.2   WHAT DATA ARE USEFUL FOR LOGISTICS?

The market for Big Data analytics in logistics was worth $3.55 billion in 2020. According to experts, it will increase to $9.28 billion by 2026 at an average annual growth rate of 17.31%. Organizations need to be able to assess the expanding amount of incoming data and enhance logistical processes. The supply chain's lack of openness and visibility is the primary issue that the IoT in transportation addresses. Managers often only have control over the transportation of commodities at the points of shipping. What happens to the shipment in between checks cannot be determined [17]. Lack of transparency makes it difficult to determine the circumstances of the products' transportation and if the carrier complies with the contract's requirements. Furthermore, there is no way to know:

1. a time-specific location of the cargo;
2. whether the products are damaged (e.g., if the container's door is closed and the packing is airtight);

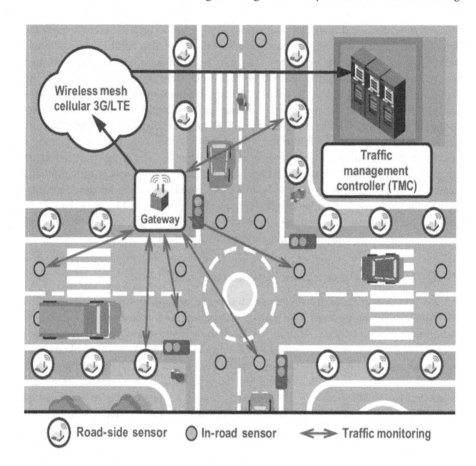

FIGURE 7.3 Automation in traffic management using IoT.

3. if the weather conditions (temperature, humidity, and other indicators) are adhered to throughout transit;
4. adverse weather conditions or traffic incidents might delay the delivery of products;
5. whether the transportation company veers from the path;
6. in addition to other concerns like whether the goods will arrive on schedule.

Delivery becomes blind when this information is absent. The management lacks knowledge of the location and solution of the supply chain issue. Improvements may be made with the correct data at hand. In addition to giving visibility into the supply chain, sensors generate a constant IoT data flow [16].

7.4.3 How a Continuous IoT Data Flow Works in Logistics?

IoT sensors are fitted to cargo-carrying packages, cartons, or containers. The location and attributes of the shipment are collected by devices. They send data to a cloud application via geographical positioning systems (GPS) or mobile networks. Members of the supply chain who have access may utilize this program. Any moment may be used to remotely control the cargo. You don't have to wait for the items to arrive at their destination to achieve this. When issues arise, the supplier takes swift action to address them or avoid them. The danger of supply chain interruption is decreased by IoT in transportation [19].

Big data in logistics and supply chain management can be used to:

- optimize routes;
- track the state of the cargo;
- evaluate the carrier's work;
- manage door-to-door delivery;
- improve warehouse operations;
- minimize damage to goods during shipment;
- predict the demand for goods, etc.

7.4.4 HOW TO IMPLEMENT THE IoT IN LOGISTICS?

Implementing the IoT in transportation takes time and strategic action:

- identify problems in the supply chain;
- set goals for the implementation of the IoT;
- develop software for Big Data in logistics and supply chain management;
- install sensors and link them to a cloud-based monitoring application;
- train employees for the transition to a new format of work.

Starting with business needs is a good idea when organizing the switch to IoT logistics. It will be simpler to develop a plan for implementing new technologies if your digitalization goals have been established. Employees will complete tasks faster and with fewer mistakes if they are prepared for changes [23]. Real IoT ecosystem development can only happen after that. Find a trustworthy technology partner is the most important factor. Business needs may be implemented, and an IoT transportation solution can be made by a team of IT experts.

7.5 SMART HEALTHCARE: A NEW WAY OF DIAGNOSIS

Quick question for all: have you ever canceled a doctor's visit due to a client meeting? Early symptom identification aids in your doctor's ability to treat the health problem with the proper medicine. Why don't you think about utilizing cutting-edge technology like Big Data, IoT, cloud computing, etc., to keep track of your present health situation?

Recently, there has been a significant increase in the global synergy between healthcare and technology. For the upcoming generation of eHealth and mHealth services, for instance, the IoT and Big Data Analytics are becoming more and more popular. Grand View Research, a market researcher, predicts that by 2022, the IoT in healthcare, which includes the markets for medical equipment, systems, software, and services, will have already reached a market size of $500 billion and is increasing exponentially [27]. The government organization realized the need for smart healthcare earlier as Information and Communication (ICT) has focused more on healthcare and has proposed more solutions that are cost-effective from a health care perspective. One such solution is the development of Electronic Health Record (EHR) systems by ICT. The whole collection of a patient's medical history, including current prescriptions, current diagnoses, vaccines, and lab results, is included in the EHR. Sharing information with the physicians using this EHR is quite useful. It could help improve the relationship between patients and doctors. Digital health care originally began with the electronic storage of patient health records, followed by patient assistance through online portals that were more adaptable and practical for both patients and physicians, also known as linked health. Wi-Fi, Bluetooth, and smartphones all play a role in making it feasible. Patients and providers may communicate instantaneously without having to physically meet [32]. For instance, a hypertension patient who visits the doctor for regular checkups will be able to transmit weekly blood pressure conditions through which the doctors can predict their health conditions and suggest medicines earlier in order to prevent the abnormal stage. In order to ensure patient conditions even at home, linked health has evolved into smart health (refer Figure 7.4), where smartphones are coupled with wearables and other devices. The primary objective of smart health (IoT in health monitoring) is to reduce hospitalization costs with prompt answers to a variety of medical issues [30]. Then, the remote location receives the microchip data that has been gathered. Data gathering devices, local gateway servers, and end systems make up the three primary parts of these systems. For instance, temperature and heartbeat sensors are utilized to gather data and transmit it to a doctor's system. The servers that may provide doctors with access

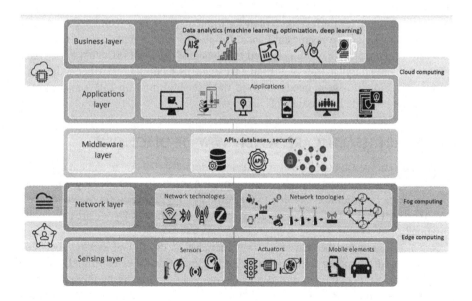

FIGURE 7.4 Smart Diagnosis using IoT application.

to patient data in real time must be continuously monitored. By analyzing medical records, data maintenance and monitoring can improve illness prediction in addition to patient health monitoring.

7.5.1   IoT in Healthcare: A Network of Healing Devices

- **Connected Medical Devices:** Wearables, medical sensors, and intelligent medical equipment make up the IoT in healthcare, which is a tapestry of linked gadgets. A constant stream of health data is provided through wearable technology, which continuously monitors vital indicators [29]. The network that is made possible by medical sensors, which are found in everything from glucose meters to pacemakers, allows for remote monitoring and prompt action.
- **Remote Patient Monitoring:** Remote patient monitoring is one of the IoT's most important functions in smart healthcare. With the ability to watch vital signs and act quickly if irregularities are found, healthcare professionals may now monitor chronically ill patients in real time [23]. This lessens the strain on medical institutions while simultaneously improving patient outcomes.
- **Enhanced Medication Management:** Patients are sure to follow their recommended prescription regimes thanks to IoT-enabled smart drug dispensers. These gadgets have the ability to remind users to take their drugs, distribute them at predetermined intervals, and even notify medical professionals if a patient deviates from the recommended course of action.

7.5.2   Big Data Analytics: The Diagnostic and Predictive Engine of Healthcare

- **Data-Driven Diagnostics:** Diagnostics is undergoing a transformation thanks to Big Data analytics, which is fueled by the enormous information produced by IoT devices. To find patterns and trends, patient data, including genetic data, lifestyle data, and real-time health measurements, are evaluated [19]. This facilitates the creation of individualized treatment regimens and early illness identification.
- **Predictive Analytics for Preventive Healthcare:** It is possible to stop illnesses before they start using predictive analytics. Healthcare professionals are able to identify and reduce possible health risks by studying past health information, lifestyle choices, and genetic predispositions. Preventive healthcare has the ability to lessen the strain on the healthcare system and enhance general population health.
- **Operational Efficiency in Healthcare Facilities:** Big Data analytics improves healthcare institutions' operational effectiveness in addition to patient care. As a result, downtime is minimized, and the smooth operation of the healthcare infrastructure is guaranteed. It also improves resource allocation, simplifies processes, and makes predictive maintenance for medical equipment possible [33].

We talked about how IoT and Big Data affect and play a vital role in healthcare sector, but the true potential of smart healthcare is realized at the intersection of IoT and Big Data. The seamless integration of real-time patient data with sophisticated analytics tools creates a healthcare ecosystem that is both proactive and personalized.

- **Plans for Individualized Therapy:** A vast amount of patient-specific data is produced by IoT devices. This data, when combined with Big Data analytics, enables healthcare practitioners to customize treatment programs according to a patient's particular health profile. The shift to individualized treatment guarantees more efficient interventions and reduces unfavorable consequences.
- **Enhancing Patient Engagement:** Smart healthcare encourages active patient participation in their own well-being. Wearable devices, patient portals, and health apps empower individuals to monitor their health, adhere to treatment plans, and stay connected with their healthcare providers. This engagement fosters a collaborative approach to healthcare [17].
- **Telemedicine and Remote Consultations:** The development of telemedicine has been aided by the combination of IoT and Big Data. Healthcare professionals may monitor patients remotely, conduct remote consultations, and undertake timely interventions thanks to real-time health data. In order to reach underserved or rural groups, this is very effective.

With the promises, there comes a set of challenges from the IoT and Big Data in smart healthcare. Data security, interoperability, and ethical considerations regarding patient privacy are paramount. Striking the right balance between harnessing the power of data and safeguarding patient confidentiality is an ongoing challenge that the healthcare industry must address [29]. But as we said with the promises with the state, the symbiotic relationship between IoT and Big Data is reshaping ecosystem. Smart healthcare is not a distant vision; it is the present reality, transforming the patient experience and redefining the contours of medical practice.

7.6 SMART GRID: SUSTAINABLE ENERGY, EFFICIENT CONSUMPTION

With smart technologies, there is an emerging smart energy source and a smart grid. The term "smart grid" refers to the energy infrastructure of the future, which combines transmissions, transformers, and substations that deliver energy to homes with cutting-edge equipment like computers, automated technology, and other new gadgets. This equipment enables digital communication or information transmission while also supplying energy to homes and businesses. The extra infrastructure layer that enables two-way communication between consumer devices and transmission lines is what the term "smart grid" alludes to. Because of the growth of various advancements, like IoT and cloud computing, this two-way communication is now feasible [28]. The smart grid is an essential component of energy because it enables energy suppliers to utilize the smart grid to its maximum potential. The new infrastructure that places a focus on linked devices is referred to as the "smart grid." A layer of communication between regional actuators, centralized controllers, and logistical units is made possible. Because it enables quicker emergency reaction times, more effective resource usage, and even better network delivery through automation, this layer of communication is helpful in a variety of contexts [31]. While maintaining connectivity between various devices, including generators and consumer electronics, is the focus of the smart grid. The smart grid is a noteworthy accomplishment since it represents a significant advancement for the energy sector. It offers a number of advantages, including improved energy transmission efficiency, cheaper management costs, greater security, fewer operational expenses, and better integration of renewable energy. Smart grid analytics are necessary to analyze the information produced since the smart grid inevitably creates a lot of data. Any value that might be derived from the data would be impossible otherwise. Connecting various items to the Internet in order to get data from them, evaluate that data, and make future choices is the core concept of the IoT. The number of devices linked to the IoT has dramatically expanded recently as a consequence of rapid technological advancements and corporate digitalization. As a result, there is a need to apply Big Data and Big Data analytics to IoT, and the volume of data has also significantly increased. Big Data and Big Data analytics offer a great deal of promise to improve decision-making by drawing out useful information from the vast amounts of data [34]. Below is a description of the primary criteria (both functional and non-functional) for Big Data and analytics in IoT.

- **Connectivity:** With the diverse items on the network, connectivity in the IoT is mostly omnipresent. In a smart environment, several things are connected to the Internet via sensors. IoT services primarily rely on machine-to-machine (M2M) communication protocols that must be able to manage several streams, and they immediately benefit from the distributed storage and processing infrastructure provided by the cloud. IoT must first and foremost deliver dependable connections for large data and analytics. Big Data and analytics will be able to effectively combine and integrate the enormous volumes of sensor data produced by machines thanks to reliable connection [34]. Many of the things in our environment can connect to the computing and high-performance infrastructure and support IoT services using modern wireless networks like Wi-Fi and 4G/5G.
- **Storage:** Massive volumes of heterogeneous data now demand a lot more storage, even on a real-time basis and on inexpensive technology. Large amounts of unstructured data must be handled by Big Data storage, and it must also offer low latency for analytics. The existence of several IoT data sources, such as sensor data, social media, and other sources, as well as the fact that they are all modeled differently and use multiple communication protocols and interfaces, presents a hurdle. Despite the fact that Big Data technology offers some IoT-efficient data storage capabilities, more durable solutions are needed.
- **Quality of Services:** Quality of service (QoS) refers to the capacity to guarantee a particular degree of performance to the data flow. The IoT

guarantees an efficient flow of data from the sources that produce large data; thus, the IoT network must be dependable and deliver this promise. Big data and analytics rely heavily on the QoS of the IoT network [26]. IoT is expanding quickly and making important moves to enhance streaming analytics and offer quick decision-making. Real-time data regarding IoT-connected items is sent and requires real-time analysis.

- **Real-Time Analytics:** IoT is expanding quickly and making important moves to enhance streaming analytics and offer quick decision-making procedures. Real-time information regarding IoT-connected items is sent and requires real-time analysis. Big data analytics executes real-time queries on the majority of the streaming data from web-enabled items to swiftly extract information, make choices, and interact with the devices and people in real time [16]. Big data employs an operational database for the streaming data.

- **Benchmark:** Many firms have started moving their operations online using IoT as a result of the rapid digitalization of enterprises. The massive volume of data linked through IoT devices is currently posing issues for many enterprises in terms of storage and analysis. It takes a thorough comprehension of the issues to find answers to those difficulties. In this context, benchmarks are crucial because they enable firms to compare the effectiveness of Big Data and analytics solutions.

Smart grid analytics will have a major influence in the future. With the developed world's energy infrastructure transitioning to a smart grid, an analytics platform capable of collecting and analyzing data from many endpoints is required. The correct analytics platform enables utility firms to more efficiently allocate resources, as seen in Figure 7.5, save costs and identify new ways to service consumers [21]. Furthermore, the correct data analytics platform enables them to make the best use of the data provided. Every analytics business, including SAS, is aiming to deliver some form of smart grid Big Data analytics, since utility companies will search for any means to enhance energy management. Let's briefly examine the benefits of data analytics.

- Collect and analyze data to improve service quality.

  Large amounts of data are generated by the smart grid as a result of IoT devices like smart meters. IoT devices are installed in various smart grid components, including substations and consumer electronics. Petabytes of data are generated by these devices, and without smart grid Big Data analytics, it is hard to make sense of the data. The data analysis capabilities of the analytics platforms may produce priceless insights that have a number of positive effects, including cost savings and improved operational effectiveness. With the help of smart grid analytics, energy businesses may quickly and efficiently solve problems with their finances and system operations. This results in various enhancements to grid optimization and client engagement [25].

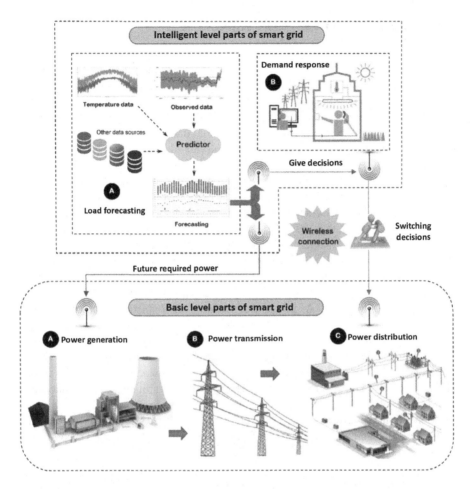

FIGURE 7.5 Efficient power generation and distribution.

- It analyses unstructured data.

  A smart grid generates a lot of unstructured data, and it can be difficult to analyze this type of data. Unstructured data must occasionally be real-time analyzed as well. Big data analytics for the smart grid is capable of analyzing unstructured data. To increase performance, uptime, and productivity, for instance, SAS Asset Performance Analytics collects data from sensors and MDM.

- Analytics comes in different formats.

  Smart grid Big Data analytics comes in different formats to suit the needs of the energy company. Utility firms can choose between point solutions and a software platform containing a suite of software solutions. Point solutions are effective because they target a specific problem. However, a single multisolution platform offers its fair share of benefits because it allows for greater flexibility and can be seen as a long-term investment [24].

7.7 SMART INVENTORY: CHANGING THE WAREHOUSE MECHANISM

A cutting-edge inventory solution, the smart inventory management system is driven by cutting-edge technology such as robotics, RFID tracking, smart shelves, IIoT (Industrial IoT), and more. To keep the best possible inventory levels, the system gathers and evaluates real-time data on your items' availability, location, condition, and shipping status. It reduces the need for manual intervention, simplifies inventory handling throughout the distribution cycle, and provides 360° warehouse visibility. Real-time visibility into inventory levels is provided by an intelligent inventory management system. Using cutting-edge technologies like IoT and AI, it tracks stock movements and forecasts demand. Businesses may streamline processes, save costs, and make intelligent judgments. Inventory management is being revolutionized, leading to improvements in efficacy, accuracy, and profitability. Effective inventory management is required rather than conventional methods for competitive advantage. With intelligent inventory management solutions that utilize AI and IoT, handling is revolutionized [31]. By implementing this approach, 65% of businesses experienced 50% fewer stockouts and better accuracy. The amount of surplus inventory was reduced by 30% thanks to AI-powered forecasting. A sophisticated software smart inventory management system aids businesses in monitoring and streamlining their procedures. Innovative technology is used by smart inventory management solutions. Contrary to manual or outdated approaches, it aids in automating and enhancing inventory-related activities. These systems keep track of stock movements and provide real-time visibility into inventory levels. It also facilitates making decisions based on data [15].

There are many different types of inventory management systems. Each kind supports the fulfillment of a certain set of organizational requirements. These are the major groups:

- **Barcode-Based Systems:** These systems use barcode scanners to track inventory items. Each item has a unique barcode, which is scanned in order to keep track of stock movements. This aids businesses in updating inventory levels [14].
- **RFID-Based Systems:** This technology uses tags and readers to track inventory items. The data contained on RFID tags is wirelessly collected by RFID readers. As a result, line-of-sight scanning is no longer required. It aids in precise and automated inventory tracking.
- **IoT-Enabled Systems:** These include the ability to track items and devices like sensors and beacons. It provides real-time analysis of supply chain movements, weather conditions, and stock levels. Through the use of IoT, enterprise software development services like Appsierra provide proactive stock management for the company [18].
- **Cloud-Based Technologies:** These apps make use of cloud servers to store stock information and manage inventory. It makes things adaptable, interoperable, and accessible from everywhere in a corporate framework.

- **AI-Driven Systems:** This system aids in the analysis of data from several sources. It also aids in controlling stock levels and identifying patterns and anticipated demand. Processes for making decisions can be computerized and automated using this technology. AI-powered technologies adapt and learn to increase the effectiveness of stock administration.

Industry leaders from all around the world have realized how smart inventory management can boost their operational effectiveness and enhance customer satisfaction. Smart inventory solutions are used by companies across a range of industry sectors to improve supply chain operations, obtain accurate insight into inventory levels, minimize reliance on human labor, and streamline inventory workflows [20]. Let's elaborate what key features a smart inventory management system holds.

- **Automation:** A smart inventory system gives users the ability to automate manual processes like item monitoring and counting, inventory database updating, report generating, restocking, etc., which saves time and effort and lowers the risk of human mistakes.
- **Real-Time Inventory Visibility:** With the help of a smart inventory system's real-time updates on item status, amount, location, and movement, managers can see more of their inventory and make wise demand forecasting choices.
- **Inventory Optimization:** A smart inventory system helps maintain the proper quantity of inventory and execute a just-in-time inventory strategy to save unneeded warehousing expenditures by spotting overstocks and stockouts.
- **Supply Chain Disruption Detection:** Real-time monitoring of all warehouse activities enables early detection and swift remediation of supply chain problems. If there are any disruption patterns, they may be rapidly examined to locate the bottlenecks and make the appropriate modifications to the stocking procedures.
- **Greater Customer Satisfaction:** Creating a great customer experience starts with smart inventory management. Prompt shipment, efficient order processing, analytics-driven planning to prevent overselling and ensure stock availability—all of these help you establish a solid reputation as a reliable business and develop enduring relationships with your customers.

The features offered by a Smart Inventory system integrated with technologies like IoT and Big Data has brought a revolution in the Business.

7.7.1 But How Does a Smart Inventory Management System Work?

Innovative technology and effective business procedures come together in a smart inventory management system, transforming the way firms manage their goods. The system's fundamental components include automation, real-time data, and cutting-edge technology, which are used to optimize the many aspects of inventory control. Through barcoding or RFID tagging, each item in the inventory is given a special identification number. These IDs make it easier to collect automatic data, which makes it possible to trace products easily as they travel through the supply chain. Then, in order to provide accessibility and updates from various places and devices, this real-time data is synchronized to a central cloud-based database. The system integrates with point-of-sale and sales systems to operate in conjunction with other corporate activities [22].

Various big business organizations realized the need and potential of this management at the earliest. For example, Decathlon, a French sports equipment retailer, deployed a smart inventory solution called StockBot in its stores across the world network to assist the shop staff. It is nothing but an inventory-tracking and data-collecting system that uses RFID tracking and smart navigation. It helps the staff in focusing more on the customers by automatically collecting the inventory data and updating it in real time while moving around the store. One more such use of technology was seen when DHL, a global logistics service provider, installed automation in their smart warehouses. There, humans and mobile and stationary robots collaborate. Static robots can choose items and handle packages of all sizes and forms using machine vision. Autonomous mobile robots may approach employees and deliver the items, lessening the physical burden on them and enhancing their health and safety [19]. In order to discover blocked routes or regions that become congested during peak hours and increase navigation efficiency, wireless tracking technology assists in tracking the movement of warehouse personnel and equipment in real time. The inventory levels are automatically updated in response to sales or other transactions, giving accurate and recent data. Automated reorder triggers boost productivity by making sure that replenishment orders are sent out right away when inventory levels drop below certain criteria. Real-time communication on stock levels and order statuses is made possible by the integration, which goes outside the company to suppliers and partners. Automated Storage and Retrieval Systems (AS/RS) are one example of an automation technology that optimizes the storage and retrieval of items in smart warehouses, increasing overall operational efficiency. These systems' capacity to offer thorough reporting and data analytics capabilities is a crucial feature. In order to decide on pricing, promotions, and inventory levels intelligently, businesses may acquire insight into stock turnover rates, spot trends, and find market niches. Finally, A sophisticated system for managing inventories has become an essential tool for companies. They may increase customer satisfaction, cut expenses, and streamline processes thanks to it. AI, IoT, and data analytics control are included. These agreements provide businesses with real-time data, automate routine tasks, and encourage proactive decision-making. Businesses may manage their stock if they have a sharp stock administration structure [27]. In the constantly changing business environment, companies and their goods can gain a competitive edge and prosper.

7.8   SMART CITIES: A NEW LIFESTYLE OF LIVING WITH EASINESS

Growing population, over-populated public transport, messed-up sewage system, traffic congestion, and many more are disabling the functioning of a city. Smart City, as we say, a high-performance urban environment where the aim is to optimize the use of resources and access to services. Smart cities use technology advancements to enhance key facets of our quality of life. The deployment of services is made possible by a number of elements, which may be categorized under the categories of IoT and Big Data. These elements include intelligent sensors, connection, access to data, and cloud applications. According to the United Nations' most recent estimate, over 68% of people will live in urban areas by 2050 [17]. In 1950, there were 751 million people living in urban areas; by 2018, that number had risen to 4.2 billion. In a study by Bibri and Krogstie, 70% of all-natural resource usage occurs in metropolitan areas, which has resulted in ecosystem degradation, pollution of the environment, and energy shortages. Limited access to resources is a major issue in the creation of these communities, since they were established with the intention of lowering prices and unemployment rates, with a special focus on tackling climate change and drinking water supplies. Therefore, there is an urgent need to adopt creative solutions to help the public deal with all of the aforementioned challenges [23]. Here, some of its requirements and features are listed.

- A robust framework that provides safe and open access;
- A citizen-oriented architecture;
- A sizable amount of wearable and mobile public and private data can be stored, found, shared, and tagged so that citizens can access information from anywhere, if necessary;
- An application with analytical and integrated features;
- A smart physical and network infrastructure that allows for the transfer of large amounts of data.

The emergence of the IoT approached the abstract concept of the smart city into reality. Superior healthcare, safety, comfort, and wisdom are among the benefits that residents of smart cities are supposed to get. In other words, they are designed to enhance quality of life. The IoT may be considered the technology most responsible for the world's most important change. Intelligent information systems that manage and regulate a smart city's resources must take the nexus of food, energy, and water into account. To address the aforementioned difficulties and the vast data generated by such systems, IoT-based solutions must be developed to ensure optimal supply and great resource efficiency [21]. Massive efforts have been made to construct smart homes, smart transportation, smart traffic management, smart trash disposal, smart energy management, smart healthcare, as well as numerous amenities, all working together to establish a smart city; refer Figure 7.6. IoT, a relatively new idea, gathers data from household appliances, public transit, mobile devices, and other sources. The entire range of everyday electronic devices found in such an environment, including household appliances such as freezers, water heaters, washing machines, microwave ovens, refrigerators, dishwashers, stoves, and air conditioners, as well as emergency alarms, wristwatches, garage doors, and vending machines, are all connected to an IoT network and can be operated remotely. The demand from citizens for innovative services that promote a high-quality living is constant and growing [35]. However, a number of obstacles arise, including the size, capacity, and availability of resources as well as the legal framework that governs and controls them. As a result,

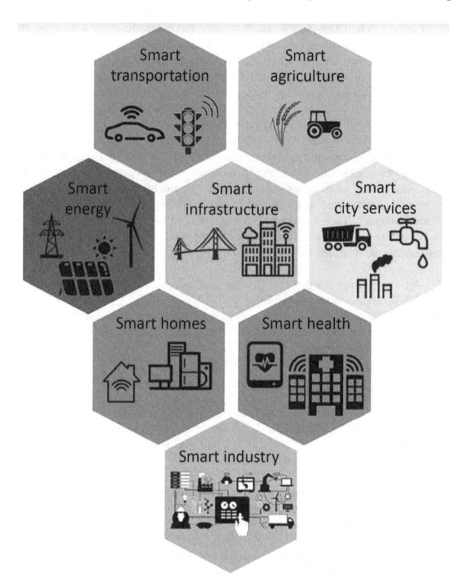

FIGURE 7.6   Smart city-improving lifestyle, reducing congestion.

Big Data is created from the data collected from several sources and produced by different IoT devices. We are surrounded by data sources wherever we look, including computers, smartphones, GPS, cameras, and even other people. The past few years have seen an acceleration in data collection due to a variety of applications, including social media websites, digital photos and videos, commercial transactions, advertising apps, games, and many more. The data being analyzed by the government and private organizations gives them insightful information that can be useful in solving day-to-day problems. Some of the benefits of a smart city are:

- **Efficient Resource Utilization:** Since many resources are becoming scarce or extremely expensive, integrating methods to have a better and more controlled usage of these resources is essential. Prioritization should be given to the use of technological solutions like enterprise resource planning and geographic information systems. With monitoring systems in place, it will be easier to locate waste areas and allocate resources more wisely, which will also aid in cost reduction and the consumption of less energy and natural resources. Since they may enhance cooperation across various apps and services, the connectivity and data collection capabilities of smart city apps are very crucial [10].
- **Better Quality of Life:** Better services, more efficient living and working environments, and less waste (of time and resources) will all contribute to a greater standard of living for citizens in smart cities. This is due to better living and working environment planning, more efficient transit networks, faster and better services, and the availability of enough information to make informed decisions.
- **Higher Level of Transparency and Openness**: The need for better management and control of the numerous smart city components and apps will lead to higher levels of openness and interoperability. Data and resource sharing will become commonplace. This will also result in increased information availability for all parties involved. This will encourage collaboration and communication among organizations, which will result in the creation of fresh apps and services that enhance the smart city even more. One example is the US government, which collected and made accessible a wide range of information, publications, and content in the spirit of openness and transparency. These made it possible for individuals and government agencies to share and use data in an effective manner [11].

A solid information and communication technology (ICT) infrastructure is needed to facilitate the deployment of Big Data applications. ICT promotes smart cities because it offers innovative solutions that would not be feasible without it. By giving them simple ways to manage their services from various areas or places, for instance, it facilitates effective transport planning and lowers transportation expenses. By providing the tools for storage and analysis, ICT, Cloud, and Big Data solutions may help solve a number of problems. Furthermore, this will help a smart city go further to the innovation stage by encouraging collaboration and communication among its numerous constituent parts. One approach to do this is to create Big Data communities that work as a single unit to promote cooperative and creative solutions for areas like industry, safety, environment, health, and education. This helps with real-time solutions to issues in crowd management, transportation, and agriculture since apps and systems are interconnected and information may move freely between apps and organizations [12]. Furthermore, coherent, trustworthy strategic plans that extend beyond isolated initiatives or projects are crucial for smart cities. Plans of this kind must take into account the several requirements of smart cities (technological, social, and physical) and refrain from addressing each component in isolation. A more comprehensive understanding of the requirements will be provided by the holistic approach, which will result in more well-rounded, well-designed solutions for smart cities as opposed to isolated, hardly detectable applications and components. Thus, Big Data and smart cities are two cutting-edge and significant ideas that, when combined, enable the creation of smart city applications that support sustainability, improved resilience, efficient governance, improved quality of life, and wise resource management [14].

7.9 CONCLUSION AND FUTURE SCOPE

Smart cities, inventory systems, healthcare, grids, and transportation have all undergone significant change as a result of the convergence of IoT and Big Data. Through better resource management and public services, these technologies have raised the standard of urban living in smart cities. Supply chains have become more efficient because of the use of smart inventory systems, which also automate inventory control and provide real-time data. Patient monitoring that is ongoing and individualized treatment programs have transformed healthcare. Reduced environmental effect is ensured by smart networks, which also assure efficient and sustainable energy delivery. IoT and Big Data are used by smart transportation systems to optimize routes, improve traffic management, and promote eco-friendly mobility options [8].

These technologies will undoubtedly improve significantly in the future. By combining 5G and edge computing, smart cities will improve connectivity, allowing for quick communication and more intelligent urban planning. Robotics, artificial intelligence, and improved predictive analytics will increase automation for smart inventory systems. Wearable technology, AI-powered diagnostics, and telemedicine will all become more prevalent in healthcare [6]. Energy storage advances, increased system resiliency, and broad use of renewable energy sources will all contribute to the evolution of smart grids. The integration of intelligent transportation networks, better traffic management, and more autonomy are all aspects of the future of smart transportation. In order to ensure responsible and inclusive deployment of IoT and Big Data technologies, ethical issues will be crucial in defining this future.

REFERENCES

1. Pantelis K, Aija L. Understanding the value of (big) data. In *2013 IEEE International Conference on Big Data.* IEEE; 2013, Silicon Valley, CA, USA. pp. 38–42.
2. Khan Z, Anjum A, Kiani SL. Cloud based big data analytics for smart future cities. In *Proceedings of the 2013 IEEE/ACM 6th International Conference on Utility and Cloud Computing.* IEEE Computer Society; 2013, Dresden, Germany. pp. 381–386.
3. Kitchin R. The real-time city? Big data and smart urbanism. *GeoJournal.* 2014;79(1):1–14.
4. Townsend AM. *Smart Cities: Big Data, Civic Hackers, and the Quest for a New Utopia.* WW Norton & Company; New York City 2013.
5. Batty M. Big data, smart cities and city planning. *Dialogues Hum Geog.* 2013;3(3):274–9.
6. Vilajosana I, Llosa J, Martinez B, Domingo-Prieto M, Angles A, Vilajosana X. Bootstrapping smart cities through a self-sustainable model based on big data flows. *IEEE Commun Mag.* 2013;51(6):128–34.

7. Michalik P, Stofa J, Zolotova I. Concept definition for Big Data architecture in the education system. In *2014 IEEE 12th International Symposium on Applied Machine Intelligence and Informatics (SAMI);* 2014, Delhi, India. pp. 331–334.
8. Fan W, Bifet A. Mining big data: Current status, and forecast to the future. *ACM SIGKDD Explor Newsl.* 2013;14(2):1–5.
9. Al-Hader M, Rodzi A. The smart city infrastructure development & monitoring. *Theor Empir Res Urban Manage.* 2009;4(2):87–94.
10. Bertot JC, Choi H. Big data and e-government: Issues, policies, and recommendations. In *Proceedings of the 14th Annual International Conference on Digital Government Research.* ACM, Quebec City, Canada; 2013. pp. 1–10.
11. Kramers A, Höjer M, Lövehagen N, Wangel J. Smart sustainable cities-Exploring ICT solutions for reduced energy use in cities. *Environ Model Software.* 2014;56:52–62.
12. Neirotti P, De Marco A, Cagliano AC, Mangano G, Scorrano F. Current trends in smart city initiatives: Some stylised facts. *Cities.* 2014;38:25–36.
13. Tantatsanawong P, Kawtrakul A, Lertwipatrakul W. Enabling future education with smart services. In *2011 Annual SRII Global Conference (SRII). IEEE;* 2011. pp. 550–556.
14. West DM. Big Data for Education: Data Mining, Data Analytics, and Web Dashboards. Governance Studies at Brookings. 2012. Available at https://www.brookings.edu/~/media/Research/Files/Papers/2012/9/04%20education%20technology%20west/04%20education%20technology%20west.pdf.
15. Marsh O, Maurov-Horvat L, Stevenson O. Big Data and Education: What's the Big Idea? UCL Policy Briefing. 2014. Available at https://www.ucl.ac.uk/public-policy/public-policy-briefings/big_data_briefing_final.pdf.
16. Aguilera G, Galan JL, Campos JC, Rodríguez P. An accelerated-time simulation for traffic flow in a smart city. *FEMTEC.* 2013;2013:26.
17. U.S. Department of Energy. Smart Grid/Department of Energy. Available at https://energy.gov/oe/technology-development/smart-grid. Retrieved September 23, 2015.
18. Yin J, Sharma P, Gorton I, Akyoli, B. Large-scale data challenges in future power grids. In *2013 IEEE 7th International Symposium on Service Oriented System Engineering (SOSE).* IEEE; San Francisco, CA, USA 2013. pp. 324–328.
19. Mohamed N, Al-Jaroodi J, Real-time big data analytics: Applications and challenges, In *2014 International Conference on High Performance Computing & Simulation (HPCS);* 2014. pp. 305–310.
20. Khan M, Uddin MF, Gupta N. Seven V's of Big Data understanding Big Data to extract value. In *2014 Zone 1 Conference of the American Society for Engineering Education (ASEE Zone 1).* IEEE; Bridgeport, Connecticut, USA 2014. pp. 1–5.
21. Su K, Li J, Fu H. Smart city and the applications. In *2011 International Conference on Electronics, Communications and Control (ICECC).* IEEE; Ningbo, China 2011. pp. 1028–1031.
22. Lee CH, Birch D, Wu C, Silva D, Tsinalis O, Li Y, Guo Y. Building a generic platform for big sensor data application. In *2013 IEEE International Conference on Big Data.* IEEE; Washington DC, USA 2013. pp. 94–102.
23. Kim GH, Trimi S, Chung JH. Big-data applications in the government sector. *Commun ACM.* 2014;57(3):78–85.
24. Chourabi H, Nam T, Walker S, Gil-Garcia JR, Mellouli S, Nahon K, Scholl HJ. Understanding smart cities: An integrative framework. In *2012 45th Hawaii International Conference on System Science (HICSS).* IEEE; Maui, USA 2012. pp. 2289–2297.
25. Xiaofeng M, Xiang C. Big data management: Concepts, techniques and challenges. *J Comput Res Dev.* 2013,1.98.

26. Borkar V, Carey MJ, Li C. Inside Big Data management: Ogres, onions, or parfaits?. In *Proceedings of the 15th International Conference on Extending Database Technology.* ACM; Berlin, Germany 2012. pp. 3–14.

27. Chaudhuri S. What next?: A half-dozen data management research goals for big data and the cloud. In *Proceedings of the 31st Symposium on Principles of Database Systems.* ACM; Pennsylvania USA 2012. pp. 1–4.

28. Dittrich J, Quiané-Ruiz JA. Efficient big data processing in Hadoop MapReduce. *Proc VLDB Endowment.* 2012;5(12):2014–5.

29. Middleton A, Solutions PDLR. *Hpcc Systems: Introduction to HPCC (High-Performance Computing Cluster). White Paper.* LexisNexis Risk Solutions; Georgia, United States 2011.

30. Alexandrov A, Bergmann R, Ewen S, Freytag JC, Hueske F, Heise A, et al. The stratosphere platform for big data analytics. *VLDB J.* 2014;23(6):939–64.

31. Biem A, Bouillet E, Feng H, Ranganathan A, Riabov A, Verscheure O, Moran C. Ibminfosphere streams for scalable, real-time, intelligent transportation services. In *Proceedings of the 2010 ACM SIGMOD International Conference on Management of Data.* ACM, Indianapolis Indiana USA; 2010. pp. 1093–1104.

32. Ji C, Li Y, Qiu W, Awada U, Li K. Big data processing in cloud computing environments. In *2012 12th International Symposium on Pervasive Systems, Algorithms and Networks (ISPAN).* IEEE; 2012. pp. 17–23.

33. Wu X, Zhu X, Wu GQ, Ding W. Data mining with big data. *IEEE Trans Knowl Data Eng.* 2014;26(1):97–107.

34. Tene O, Polonetsky J. Big data for all: Privacy and user control in the age of analytics. *Nw J Tech Intell Prop.* 2012;11:xxvii.

35. Business Analytics from Basics to Value. Gartner. Published on June 10, 2014. Available at https://www.slideshare.net/sucesuminas/business-analytics-from-basics-to-value. Retrieved May 4, 2015.

8 Big Data Architecture for IoT

Ravi Kumar Barwal, Ashok, and Neeraj Rohilla

8.1 INTRODUCTION

The IoT is transforming the way we interact with the world around us. In this interconnected landscape, sensors, actuators, and devices are continually collecting data, creating an ever-expanding digital footprint. This data, often referred to as Big Data, contains valuable insights that can revolutionize industries, enhance services, and fuel innovation. However, the sheer scale, velocity, and variety of IoT-generated data present significant challenges when it comes to storage, processing, and analysis. To unlock the potential of this data-driven revolution, a robust and efficient Big Data architecture for IoT is essential.

This paper is a comprehensive exploration of the intricacies of IoT data management. We will guide you through the fundamental components of building a resilient architecture that can handle the deluge of data from IoT devices. Whether you are an engineer, data scientist, business leader, or technology enthusiast, this document aims to provide you with the knowledge and insights needed to make informed decisions about architecting and implementing a Big Data solution for IoT. We will delve into data ingestion, processing, storage, analytics, security, and scalability, among other crucial considerations.

8.2 LITERATURE REVIEWS

Sr. No.	Paper Title	Year	Techniques	Findings
1.	"An Efficient Big Data Architecture for IoT Applications"	2021	Big Data architecture for IoT	Proposed an efficient architecture for IoT applications
2.	"Big Data Analytics in IoT: A Review"	2020	Various Big Data analytics techniques	Overview of IoT Big Data analytics challenges and trends
3.	"A Survey of Big Data Architectures and Machine Learning Algorithms in Healthcare"	2020	Machine learning algorithms and big data architectures	Exploration of big data architectures and machine learning methods in healthcare IoT applications

(Continued)

DOI: 10.1201/9781032673479-10

(Continued)

Sr. No.	Paper Title	Year	Techniques	Findings
4.	"A Survey on Data Management in IoT: A Smart Cities Perspective"	2020	Data management methods and techniques in IoT	Survey of data management methods and techniques in IoT, particularly in smart cities
5.	"A Comprehensive Survey on IoT Toward 5G Wireless Systems"	2018	Various IoT and Big Data algorithms and techniques	Examination of IoT and 5G convergence, highlighting Big Data's role
6.	"Big Data Analytics for IoT-Based Smart Transportation Systems: A Survey"	2017	Big Data analytics techniques for smart transportation	Survey of Big Data analytics techniques in IoT for smart transportation
7.	"Big Data Analytics in the Cloud for Internet of Things"	2016	Cloud-based Big Data analytics techniques	Discussion of Big Data analytics techniques in IoT with a focus on cloud-based solutions

8.3   ARCHITECTURE

The Big Data Architecture for IoT comprises several key components:

1. **IoT Devices:**
   - These are the sensors, actuators, and devices that collect data.
   - Devices can vary from simple sensors in a home automation system to complex industrial machines in a manufacturing plant.
2. **Data Ingestion Layer:**
   - **Message Queues**: Incoming data from IoT devices is first sent to message queues (e.g., Apache Kafka, RabbitMQ).
   - Message queues help in buffering and managing the data flow, ensuring reliability and handling data spikes.
3. **Data Preprocessing:**
   - Data may go through initial preprocessing steps to remove noise, validate, and enrich it.
   - Preprocessing may include data cleansing, transformation, and aggregation.
4. **Data Storage Layer:**
   - **Data Lake**: Raw and unprocessed data is stored in a data lake.
   - Data lakes are scalable and cost-effective storage solutions for massive volumes of IoT data.
5. **Data Warehouses (Optional):**
   - Structured or semi-structured data may be moved to data warehouses (e.g., Redshift, BigQuery) for efficient querying and analytics.

6. **Data Processing Layer:**
   - **Batch Processing**: Historical data can be processed in batches using tools like Apache Spark or Hadoop. This helps in uncovering long-term insights and trends.
   - **Stream Processing**: Real-time data analysis is performed using stream processing frameworks (e.g., Apache Kafka Streams and Apache Flink). Stream processing allows immediate insights and actions based on incoming data.
7. **Data Analytics Layer:**
   - Machine learning and AI models are applied to gain insights, predict trends, detect anomalies, and optimize operations.
8. **Data Visualization:**
   - Tools like Tableau, Power BI, or custom dashboards are used to create meaningful visualizations for end-users and decision-makers (Figure 8.1).

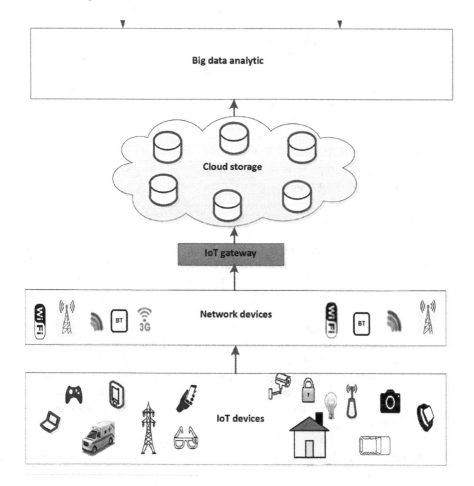

FIGURE 8.1 Big Data Architecture for IoT.

9. **Scalability and High Availability:**
   - Ensure the architecture is scalable to handle increasing data volumes and provides high availability to prevent downtime.
10. **IoT Device Management:**
   - A device management system is essential for monitoring and controlling IoT devices remotely, performing firmware updates, and managing device lifecycles.
11. **Edge Computing (Optional):**
   - Deploy edge computing resources close to IoT devices for initial data processing and reducing latency for critical applications.
12. **Feedback Loop:**
   - Continuously monitor system performance and gather feedback to improve data quality, processing speed, and overall architecture.
13. **Cost Optimization:**
   - Optimize costs by using a mix of cloud and on-premises resources, leveraging serverless computing, and regularly reviewing and adjusting resources based on usage patterns.
14. **Backup and Disaster Recovery:**
   - Implement backup and disaster recovery strategies to ensure data resilience in case of system failures or disasters.
15. **Compliance and Regulations:**
   - Stay compliant with industry-specific regulations and standards relevant to your IoT data, such as data encryption, access controls, and audit trails.
16. **Monitoring and Alerting:**
   - Implement comprehensive monitoring and alerting systems to proactively identify and address issues in real time.

8.4 APPLICATIONS

Big Data architecture for IoT applications is designed to handle the massive volumes of data generated by IoT devices and to enable various use cases and applications. This Big Data Architecture for IoT finds applications in various domains, including:

1. **Smart Home Automation:**
   - **IoT Devices**: Sensors and actuators in a smart home, such as thermostats, cameras, and lighting controls.
   - **Architecture**: Data from these devices is ingested into the system, pre-processed, and stored in data lakes. Real-time analytics can be applied for immediate responses, like adjusting thermostat settings or sending alerts for security breaches.
2. **Industrial IoT (IIoT):**
   - **IoT Devices**: Industrial sensors and machines in manufacturing plants.

- **Architecture**: Data is collected, pre-processed, and stored for batch processing. Predictive maintenance models in the data analytics layer can predict equipment failures, reducing downtime and increasing efficiency.

3. **Healthcare Monitoring:**
   - **IoT Devices**: Wearable health monitors and medical sensors.
   - **Architecture**: Data from wearable is continuously streamed for real-time analysis. Machine learning models can identify anomalies in vital signs, and alerts can be generated for medical professionals or patients.

4. **Smart Cities:**
   - **IoT Devices**: Sensors on traffic lights, waste bins, environmental monitoring stations, and public transportation.
   - **Architecture**: Data is collected and processed in real time to optimize traffic flow, reduce energy consumption, and improve public services. Visualization tools can help city officials make data-driven decisions.

5. **Precision Agriculture:**
   - **IoT Devices**: Soil sensors, drones, and weather stations.
   - **Architecture**: Data from these devices is collected and analyzed to optimize irrigation, fertilizer use, and planting schedules. This can improve crop yields and reduce resource waste.

6. **Supply Chain Management:**
   - **IoT Devices**: RFID tags, GPS trackers, and temperature sensors.
   - **Architecture**: Data from the supply chain is tracked in real time. Analytics can predict delivery times, optimize routes, and ensure product quality by monitoring temperature-sensitive goods.

7. **Energy Management:**
   - **IoT Devices**: Grid sensors, Smart meters, and energy management systems.
   - **Architecture**: Data is collected and analyzed to optimize energy distribution, detect faults, and reduce energy consumption during peak times. This can improve grid reliability and reduce costs.

8. **Fleet Management:**
   - **IoT Devices**: GPS trackers, vehicle sensors, and telematics systems.
   - **Architecture**: Data is continuously collected and processed to track vehicle locations, monitor driver behavior, and optimize routes for efficient logistics and cost savings.

9. **Environmental Monitoring:**
   - **IoT Devices**: Sensors for air quality, water quality, and wildlife tracking.
   - **Architecture**: Real-time data from sensors is analyzed to monitor environmental conditions, detect pollution events, and protect ecosystems.

10. **Retail and Customer Analytics:**
    - **IoT Devices**: In-store beacons, RFID tags, and video cameras.
    - **Architecture**: Customer behavior data is collected and analyzed to optimize store layouts, improve customer experiences, and personalize marketing campaigns.

8.5 CHALLENGES

Designing and implementing Big Data architecture for IoT presents several significant challenges, given the complexity, velocity, volume, and variety of data generated by IoT devices. The key challenges of Big Data for IoT are:

1. **Data Volume:**
   IoT devices generate massive amounts of data, which can quickly overwhelm storage and processing systems. Managing the storage, retrieval, and processing of this data efficiently is a significant challenge.

2. **Data Velocity:**
   IoT data arrives in real time or near real time, making it crucial to have systems capable of handling high data velocities. Stream processing and low-latency analytics are essential.

3. **Data Variety**:
   IoT data comes in various formats, including structured, semi-structured, and unstructured data. Handling this diverse data effectively requires versatile data processing capabilities.

4. **Data Quality:**
   IoT data can be noisy, incomplete, or inaccurate due to sensor malfunctions or environmental factors. Data cleansing and validation are essential to ensure data quality.

5. **Data Security:**
   IoT devices often collect sensitive information. Securing data both in transit and at rest is challenging, especially when dealing with a vast number of devices.

6. **Scalability:**
   As the number of IoT devices increases, the architecture must be highly scalable to accommodate growing data volumes and device connections without sacrificing performance.

7. **Latency:**
   For real-time applications like autonomous vehicles or industrial automation, minimizing data processing latency is critical. Edge computing and optimized data pipelines are necessary.

8. **Complex Event Processing:**
   Identifying and acting on meaningful events in real-time data streams can be challenging. Implementing complex event processing algorithms is necessary for timely decision-making.

9. **Privacy and Compliance:**
   IoT data may contain personally identifiable information. Complying with data privacy regulations, such as GDPR or HIPAA, while still utilizing the data, requires careful planning.

10. **Energy Efficiency:**
    IoT devices often have limited power resources. Efficient data transmission, device management, and processing are vital to extend device battery life.

11. **Data Governance:**

Establishing data governance policies and procedures for IoT data, including data retention, access controls, and audit trails, is essential for data management and compliance.

12. **Cost Management:**

Running a large-scale IoT data architecture can be costly, especially when using cloud resources. Optimizing costs while maintaining performance is a continuous challenge.

13. **Reliability and Availability:**

Ensuring high availability and fault tolerance is crucial for mission-critical IoT applications. Redundancy and disaster recovery planning are essential.

14. **Ecosystem Complexity:**

IoT ecosystems often involve multiple stakeholders, including device manufacturers, network providers, and data consumers. Coordinating and aligning interests can be challenging.

8.6 CONCLUSION

In conclusion, a meticulously designed Big Data Architecture tailored for IoT applications stands as the bedrock of our ability to navigate the complex landscape of data generated by IoT devices. With its real-time processing capabilities, scalability, and flexibility, this architecture empowers organizations to harness the wealth of information flowing from IoT sensors. Moreover, it ensures data governance and security, paving the way for responsible data management in compliance with evolving regulations. The challenge of interoperability is met with seamless integration, while latency is minimized through innovative solutions such as edge computing. Overall, this architecture not only addresses the intricacies of handling IoT data but also opens the door to a future where data-driven insights from IoT applications drive innovation, efficiency, and informed decision-making across diverse sectors.

BIBLIOGRAPHY

1. A. Nauman, Y. A. Qadri, M. Amjad, Y. B. Zikria, M. K. Afzal and S. W. Kim, "Multimedia Internet of Things: A Comprehensive Survey," *IEEE Access*, vol. 8, pp. 8202–8250, 2020, doi:10.1109/ACCESS.2020.2964280.
2. M. Stoyanova, Y. Nikoloudakis, S. Panagiotakis, E. Pallis and E. K. Markakis, "A Survey on the Internet of Things (IoT) Forensics: Challenges, Approaches, and Open Issues," *IEEE Communications Surveys & Tutorials*, vol. 22, no. 2, pp. 1191–1221, 2020, doi:10.1109/COMST.2019.2962586.
3. D. Fawzy, S. Moussa and N. Badr, "WFEC: Wind Farms Economic Classifier Using Big Data Analytics," *2017 Eighth International Conference on Intelligent Computing and Information Systems (ICICIS)*, Cairo, Egypt, 2017, pp. 154–159, doi:10.1109/INTELCIS.2017.8260046.
4. X. Larrucea, A. Combelles, J. Favaro and K. Taneja, "Software Engineering for the Internet of Things," *IEEE Software*, vol. 34, no. 1, pp. 24–28, 2017, doi:10.1109/MS.2017.28.

5. A. Bröring A., Schmid, S., Schindhelm, C.K., Khelil, A et al., "Enabling IoT Ecosystems through Platform Interoperability," *IEEE Software*, vol. 34, no. 1, pp. 54–61, 2017, doi:10.1109/MS.2017.2.

6. N. H. Morin and F. Fleurey, "Model-Based Software Engineering to Tame the IoT Jungle," *IEEE Software*, vol. 34, no. 1, pp. 30–36, 2017, doi:10.1109/MS.2017.11.

7. A. Taivalsaari and T. Mikkonen, "A Roadmap to the Programmable World: Software Challenges in the IoT Era," *IEEE Software*, vol. 34, no. 1, pp. 72–80, 2017, doi:10.1109/MS.2017.26.

8. Y. Mehmood, F. Ahmad, I. Yaqoob, A. Adnane, M. Imran and S. Guizani, "Internet-of-Things-Based Smart Cities: Recent Advances and Challenges," *IEEE Communications Magazine*, vol. 55, no. 9, pp. 16–24, 2017, doi:10.1109/MCOM.2017.1600514.

9. M. Mohammadi, A. Al-Fuqaha, S. Sorour and M. Guizani, "Deep Learning for IoT Big Data and Streaming Analytics: A Survey," *IEEE Communications Surveys & Tutorials*, vol. 20, no. 4, pp. 2923–2960, 2018, doi:10.1109/COMST.2018.2844341.

10. M. Marjani, M., Nasaruddin, F., Gani, A., Karim, A., et al., "Big IoT Data Analytics: Architecture, Opportunities, and Open Research Challenges," *IEEE Access*, vol. 5, pp. 5247–5261, 2017, doi:10.1109/ACCESS.2017.2689040.

11. O. B. Sezer, E. Dogdu and A. M. Ozbayoglu, "Context-Aware Computing, Learning, and Big Data in Internet of Things: A Survey," *IEEE Internet of Things Journal*, vol. 5, no. 1, pp. 1–27, 2018, doi:10.1109/JIOT.2017.2773600.

12. F. Alam, R. Mehmood, I. Katib, N. N. Albogami and A. Albeshri, "Data Fusion and IoT for Smart Ubiquitous Environments: A Survey," *IEEE Access*, vol. 5, pp. 9533–9554, 2017, doi:10.1109/ACCESS.2017.2697839.

13. A. Ullah, M. Azeem, H. Ashraf, A. A. Alaboudi, M. Humayun and N. Jhanjhi, "Secure Healthcare Data Aggregation and Transmission in IoT-A Survey," *IEEE Access*, vol. 9, pp. 16849–16865, 2021, doi:10.1109/ACCESS.2021.3052850.

14. W. Yu, W., Liang, F., He, X., Hatcher, et al., "A Survey on the Edge Computing for the Internet of Things," *IEEE Access*, vol. 6, pp. 6900–6919, 2018, doi:10.1109/ACCESS.2017.2778504.

15. E. M. Abou-Nassar, A. M. Iliyasu, P. M. El-Kafrawy, O. -Y. Song, A. K. Bashir and A. A. A. El-Latif, "DITrust Chain: Towards Blockchain-Based Trust Models for Sustainable Healthcare IoT Systems," *IEEE Access*, vol. 8, pp. 111223–111238, 2020, doi:10.1109/ACCESS.2020.2999468.

16. M. Chen, Y. Zhang, M. Qiu, N. Guizani and Y. Hao, "SPHA: Smart Personal Health Advisor Based on Deep Analytics," *IEEE Communications Magazine*, vol. 56, no. 3, pp. 164–169, 2018, doi:10.1109/MCOM.2018.1700274.

17. O. Elijah, T. A. Rahman, I. Orikumhi, C. Y. Leow and M. N. Hindia, "An Overview of Internet of Things (IoT) and Data Analytics in Agriculture: Benefits and Challenges," *IEEE Internet of Things Journal*, vol. 5, no. 5, pp. 3758–3773, 2018, doi:10.1109/JIOT.2018.2844296.

18. S. Rajeswari, K. Suthendran and K. Rajakumar, "A Smart Agricultural Model by Integrating IoT, Mobile and Cloud-Based Big Data Analytics," *2017 International Conference on Intelligent Computing and Control (I2C2), Coimbatore*, India, 2017, pp. 1–5, doi:10.1109/I2C2.2017.8321902.

19. M. Roopaei, P. Rad and K.-K. R. Choo, "Cloud of Things in Smart Agriculture: Intelligent Irrigation Monitoring by Thermal Imaging," *IEEE Cloud Computing*, vol. 4, no. 1, pp. 10–15, 2017, doi:10.1109/MCC.2017.5.

20. M. A. Rahman, M. M. Rashid, M. S. Hossain, E. Hassanain, M. F. Alhamid and M. Guizani, "Blockchain and IoT-Based Cognitive Edge Framework for Sharing Economy Services in a Smart City," *IEEE Access*, vol. 7, pp. 18611–18621, 2019, doi:10.1109/ACCESS.2019.2896065.

21. F. Li, H. Li, C. Wang, K. Ren and E. Bertino, "Guest Editorial Special Issue on Security and Privacy Protection for Big Data and IoT," *IEEE Internet of Things Journal*, vol. 6, no. 2, pp. 1446–1449, 2019, doi:10.1109/JIOT.2019.2908460.

22. S. Sharma, K. Chen and A. Sheth, "Toward Practical Privacy-Preserving Analytics for IoT and Cloud-Based Healthcare Systems," *IEEE Internet Computing*, vol. 22, no. 2, pp. 42–51, 2018, doi:10.1109/MIC.2018.112102519.

23. R. Lu, K. Heung, A. H. Lashkari and A. A. Ghorbani, "A Lightweight Privacy-Preserving Data Aggregation Scheme for Fog Computing-Enhanced IoT," *IEEE Access*, vol. 5, pp. 3302–3312, 2017, doi:10.1109/ACCESS.2017.2677520.

24. M. Ortiz, D. Hussein, S. Park, S. N. Han and N. Crespi, "The Cluster Between Internet of Things and Social Networks: Review and Research Challenges," *IEEE Internet of Things Journal*, vol. 1, no. 3, pp. 206–215, 2014, doi:10.1109/JIOT.2014.2318835.

25. A. Ahmad, M. M. Rathore, A. Paul and S. Rho, "Defining Human Behaviors Using Big Data Analytics in Social Internet of Things," *2016 IEEE 30th International Conference on Advanced Information Networking and Applications (AINA)*, Crans-Montana, Switzerland, 2016, pp. 1101–1107, doi:10.1109/AINA.2016.104.

26. K. Yang, S. Liu, L. Cai, Y. Yilmaz, P. -Y. Chen and A. Walid, "Guest Editorial Special Issue on AI Enabled Cognitive Communication and Networking for IoT," *IEEE Internet of Things Journal*, vol. 6, no. 2, pp. 1906–1910, 2019, doi:10.1109/JIOT.2019.2908443.

27. M. M. Rathore, A. Ahmad, A. Paul and G. Jeon, "Efficient Graph-Oriented Smart Transportation Using Internet of Things Generated Big Data," *2015 11th International Conference on Signal-Image Technology & Internet-Based Systems (SITIS)*, Bangkok, Thailand, 2015, pp. 512–519, doi:10.1109/SITIS.2015.121.

28. T. S. J. Darwish and K. Abu Bakar, "Fog Based Intelligent Transportation Big Data Analytics in The Internet of Vehicles Environment: Motivations, Architecture, Challenges, and Critical Issues," *IEEE Access*, vol. 6, pp. 15679–15701, 2018, doi:10.1109/ACCESS.2018.2815989.

29. W. He, G. Yan and L. D. Xu, "Developing Vehicular Data Cloud Services in the IoT Environment," *IEEE Transactions on Industrial Informatics*, vol. 10, no. 2, pp. 1587–1595, 2014, doi:10.1109/TII.2014.2299233.

30. P. Žarko, K. Pripužić, M. Serrano and M. Hauswirth, "IoT Data Management Methods and Optimisation Algorithms for Mobile Publish/Subscribe Services in Cloud Environments," *2014 European Conference on Networks and Communications (EuCNC)*, Bologna, Italy, 2014, pp. 1–5, doi:10.1109/EuCNC.2014.6882657.

31. C. Doukas and F. Antonelli, "COMPOSE: Building Smart & Context-Aware Mobile Applications Utilizing IoT Technologies," *Global Information Infrastructure Symposium - GIIS 2013*, Trento, Italy, 2013, pp. 1–6, doi:10.1109/GIIS.2013.6684373.

32. Y. Simmhan, Y., Aman, S., Kumbhare, A., Liu, R., et al., "Cloud-Based Software Platform for Big Data Analytics in Smart Grids," *Computing in Science & Engineering*, vol. 15, no. 4, pp. 38–47, 2013, doi:10.1109/MCSE.2013.39.

33. K. Wang, Wang, Y., Hu, X., Sun, Y., Deng, D.J., et al., "Wireless Big Data Computing in Smart Grid," *IEEE Wireless Communications*, vol. 24, no. 2, pp. 58–64, 2017, doi:10.1109/MWC.2017.1600256WC.

34. H. N. Akouemo and R. J. Povinelli, "Data Improving in Time Series Using ARX and ANN Models," *IEEE Transactions on Power Systems*, vol. 32, no. 5, pp. 3352–3359, 2017, doi:10.1109/TPWRS.2017.2656939.

35. S. Kazmi, N. Javaid, M. J. Mughal, M. Akbar, S. H. Ahmed and N. Alrajeh, "Towards Optimization of Metaheuristic Algorithms for IoT Enabled Smart Homes Targeting Balanced Demand and Supply of Energy," *IEEE Access*, vol. 7, pp. 24267–24281, 2019, doi:10.1109/ACCESS.2017.2763624.

36. R. Al-Ali, I. A. Zualkernan, M. Rashid, R. Gupta and M. Alikarar, "A Smart Home Energy Management System Using IoT and Big Data Analytics Approach," *IEEE Transactions on Consumer Electronics*, vol. 63, no. 4, pp. 426–434, 2017, doi:10.1109/TCE.2017.015014.

37. S. A. Shah, D. Z. Seker, S. Hameed and D. Draheim, "The Rising Role of Big Data Analytics and IoT in Disaster Management: Recent Advances, Taxonomy and Prospects," *IEEE Access*, vol. 7, pp. 54595–54614, 2019, doi:10.1109/ACCESS.2019.2913340.

38. H. Gibson, S. Andrews, K. Domdouzis, L. Hirsch and B. Akhgar, "Combining Big Social Media Data and FCA for Crisis Response," *2014 IEEE/ACM 7th International Conference on Utility and Cloud Computing*, London, UK, 2014, pp. 690–695, doi:10.1109/UCC.2014.112.

39. H. Zhang, M. Babar, M. U. Tariq, M. A. Jan, V. G. Menon and X. Li, "SafeCity: Toward Safe and Secured Data Management Design for IoT-Enabled Smart City Planning," *IEEE Access*, vol. 8, pp. 145256–145267, 2020, doi:10.1109/ACCESS.2020.3014622.

40. R. K. Barwal, N. Raheja, M. Bhiyana and D. Rani, "Machine Learning-Based Hybrid Recommendation (SVOF-KNN) Model For Breast Cancer Coimbra Dataset Diagnosis," *International Journal on Recent and Innovation Trends in Computing and Communication*, vol. 11, no. 1s, pp. 23–42, 2023, doi:10.17762/ijritcc.v11i1s.5991.

41. Q.-T. Doan, A. S. M. Kayes, W. Rahayu and K. Nguyen, "Integration of IoT Streaming Data with Efficient Indexing and Storage Optimization," *IEEE Access*, vol. 8, pp. 47456–47467, 2020, doi:10.1109/ACCESS.2020.2980006.

42. N. Medhat, S. M. Moussa, N. L. Badr and M. F. Tolba, "A Framework for Continuous Regression and Integration Testing in IoT Systems Based on Deep Learning and Search-Based Techniques," *IEEE Access*, vol. 8, pp. 215716–215726, 2020, doi:10.1109/ACCESS.2020.3039931.

43. F. Bock, C. Sippl, A. Heinz, C. Lauer and R. German, "Advantageous Usage of Textual Domain-Specific Languages for Scenario-Driven Development of Automated Driving Functions," *2019 IEEE International Systems Conference (SysCon)*, Orlando, FL, 2019, pp. 1–8, doi:10.1109/SYSCON.2019.8836912.

Section 3

Algorithms

9 Enhanced Cryptography and Communication Scheme Using Dynamic Cipher for IoT

Navneet Verma

9.1 INTRODUCTION

IoT is a rising technology built on top of the Internet, combining radio-frequency identification (RFID) technology, wireless communications technology, the evolved packet core standard, and other technologies [1,2]. In IoT, everything is exchanged on a global platform with real-time information; therefore, IoT security issues are increasingly obvious as it is developed. Internet is not secure, and it may get worse, giving cyber criminals plenty of room to operate and opportunities throughout the whole wireless network [3]. Many ideas and important technologies should be breakthroughs because IoT research and implementation are still in their early phases [4]. Most recent network security structures can provide certain security methods for IoT security, including authentication and encryption systems. However, it has to be modified to fit the IoT properties. Today, about two billion individuals use the Internet for a variety of activities, such as social networking, email forwarding and receiving, consuming multimedia information and services, playing games, and more. In the near future, there will be yet another important development in the usage of the Internet as a universal podium for allowing technology and smart things to connect, calculate, and administer. As a result of such a widespread information and communication infrastructure, more and more people will have access to it. The outcome will be, the information and communication industry comes with new prospects by opening the door to new services and applications that will create a bridge between the physical and digital worlds.

A trustworthy variable key authentication mechanism relying on the request-reply method is the dynamic variable cipher security certificate suggested in this study [5]. It is distinctive for its "one-time one-cipher" real-time timestamp performance, very strong data storage, and little computing overhead. It can be extensively utilized in the near field communications authentication procedure.

This chapter will cover a range of IoT security concerns and how cryptographic services may address them. The following is the order of the chapter: After the introduction, various existing security frameworks at the sensor layer are discussed in the second segment, that is, literature work. The third part largely explains the Dynamic

Cipher design concept and how its security certificate is authenticated. The usage of a dynamic variable cipher security certificate is covered in the fourth phase, which is followed by an analysis of the test findings. The conclusion of this chapter is in the fifth part.

9.2 LITERATURE WORK

In this research [6], "Elliptic-Curve Diffie-Hellman (ECDH) and QUARK hash (qh)" security algorithms are used in a novel protocol for key exchange and network joining. By adopting an effective methodological approach that emphasizes security above performance, the major focus is on deploying an efficient and high-security mechanism to decrease the strain on WSN resources. Users (patients and providers) may feel more safe knowing that this protocol is protecting their information and data, which offers a trusted key exchange. A significant security protocol flaw would exist if a security mechanism for key exchange was provided without regard for protocol performance. The protocol was put to the test against many malicious key exchange attacks. The analysis findings show that procedure is an effective defense against these dangers. When compared to the current search results, this protocol also offers an unbiased output in terms of processing and transmission expenses.

Drawbacks: Multiple users can utilize similar components of the same session key if the network is big. Further, the suggestion may establish a session key for a certain user exclusive of establishing a session key for the sensors and the other way round, if there is a hardware breakdown of the network apparatus or the suggestion has not been validated beside a variety of threats and assaults. In addition, since the base station verification processes are so crucial to the key exchange phase, any base station device malfunction might result in network disruption.

The combination of the Cloud with IoT, or the Cloud IoT paradigm, is the main topic of discussion in this article [1]. The author of this article offers a literature review on the integration of the cloud and IoT to close this gap. Here, the IoT and cloud computing fundamentals are first examined, followed by a discussion on the current factors influencing their integration. To determine the complementary features of the Cloud and IoT as well as the key factors influencing their integration into a particular context, the primary research problem for each of the new applications made possible by the adoption of the Cloud IoT paradigm is to be interesting.

Drawbacks: The disadvantages of integrating Cloud IoT include security and isolation, heterogeneity, consistency, large-scale performance, legal and social aspects, big data, monitoring via IoT sensors, and trustworthy fog computing.

"Decision parameter"-oriented techniques are widely employed; however, they frequently take centralized access control into account. In this article [7], the importance and need of movement management in intelligent mutual ecosystems and outline essential traits and limits of several existing access control methods are highlighted. After that, an expanded activity control (ACON) architecture to deal with the requirements of dynamic Smart and mutual computing systems (SCSs) is provided. Additionally, a comparison is made between the design principles for ACON for intelligent and collaborative computing systems and the access control design concepts now in use. In this research, an improved framework for next-generation ACON that takes into account group controls for practice, examination, and organized activities carried out by several users in collaborative and smart systems is provided. Their design concepts for activity management include generalization, controllability, confinement, mechanization, responsibility, and searchability. They also created the ACON framework. We think that for cutting-edge access and activity control systems, the suggested ACON architecture and design principles are essential.

In this work [2], researchers addressed security concerns in heterogeneous IoT networks using efficient elliptic curve cryptography techniques in this report. They combined the cryptographic algorithms for "NXP/Jennic 5148- and MSP430-based" IoT equipment to create a cutting-edge key negotiation method. The recommended approach also incorporates the conviction-enhancing protocol for manufacturer authentication of newly discovered devices and sensors.

Drawback: This protocol message exchange's participants are open to a man-in-the-middle attack.

The Internet of Things (IoT), which is made possible by the most recent advancements in RFID, smart sensors, transmission technologies, and Internet protocols, is described in this study [3]. The current rise in machine-to-machine (M2M), mobile, and Internet technologies can be seen as the start of the IoT. One of the key IoT technologies used in the healthcare sector is RFID technology [6]. Some of the security requirements that an RFID authentication technique should meet were acknowledged. Only three of the most current ECC-based RFID authentication approaches that have been suggested have been found to guarantee all security requirements.

Drawbacks: Collision Problems, Security and Privacy Concerns, cost, design, and integration.

According to this research [4], the fulfillment of security and privacy criteria will be crucial in the future. These specifications include data security and authentication, IoT network access management, privacy and confidence among users and devices, and the implementation of security and privacy laws. IoT technologies differ from traditional security countermeasures in terms of the communication stacks and standards used. This research paper will help to suggest the road ahead, which will allow for the enormous deployment of IoT systems in the real world.

Drawback: Authorized access permission, scalable IoT architecture, handling the huge amount of transmitted data, identification of entities.

In this paper [8], security is discussed about the healthcare industry's transformation from version 1.0 to version 4.0, where version 1.0 focused on doctors and version 2.0 interchanged hardcopies with digital records. Healthcare 3.0 is patient-centered, but Healthcare 4.0 includes telemedicine, fog computing, cloud computing, IoT, and other technologies to exchange data across different collaborators. However, developing a safe method for Healthcare 4.0 has been hard all the time. Healthcare 4.0 may use an insecure method that exposes patient email accounts, messages, and reports to hackers, resulting in a healthcare data breach. Contrarily, a secure Healthcare 4.0 method may satisfy all parties involved, including patients and upholders. This study gives a thorough examination of the state-of-the-art ideas to sustain security and privacy in Healthcare 4.0, which is inspired by the facts discussed in this paper.

Even a blockchain-based solution was also suggested. It encompasses every facet of healthcare, including processing, Machine Learning, IoT, Telehealthcare, and policy-based, network traffic, and verification systems.

Drawbacks: User authentication, ethical challenges, confidentiality and integrity, data possession, data protection policies, misuse of health records, etc.

The IoT is an archetype [9] where common things may be given identifying, sensing, networking, and processing abilities that will permit them to connect with other equipment and services through the Internet to attain several goals. To facilitate M2M communication like that visualized for the semantic web, the IoT draws on already-developed technologies like RFID and Wireless Sensor Networks as well as standards and protocols. One unanswered issue is whether or not the IoT will prove to be a viable technology if it will fizzle out or whether it will serve as a stepping stone to a different paradigm. In the end, only time will be able to provide an answer. However, the IoT can change our world by freshly fusing already existing technology.

Drawbacks: Security, Privacy, Legal/Accountability, and General.

To effectively and continually gather patient data, healthcare organizations need innovative technologies. Performance and security efficiency are still issues with traditional systems. By utilizing safe and portable signature techniques, the researcher in this study [10] suggested a "Reliable and Efficient Integrity Scheme for Data Collection in HWSN (REISCH)" to address these issues. The outcomes of the performance research show that this plan offers extremely effective data integration between servers and sensors. The security protocols are also validated using the AVISPA (Automated Validation of Internet Security Protocols and Applications) system. Furthermore, the outcomes of the security research demonstrate that REISCH is impervious to assaults according to the threat model.

Drawbacks: This scheme overcomes only security and effectiveness issues.

9.3   LAYERWISE SECURITY IN IOT

9.3.1   Current Security Structure

IoT may be summarized in terms of the following three features. The first benefit is that nodes can easily communicate with one another since everything in the world is connected to the Internet. Second, all-encompassing sensing allows for the automated identification of any IoT object. The second and third categories are the essential elements of the IoT. The third category, smart processing, is differentiated by computerization, self-feedback, intelligent control, etc. Three levels, comprising the Application Layer Security, the Network Layer Security, and the sensor layer security, may be seen in the IoT security system as a whole. Secure data sensing, dependable data conveyance, and safe data control are at the heart of IoT security, as shown in Figure 9.1 of the schematic.

The security mechanism of the IoT must be developed based on the major technologies that may be implemented in each layer and the security risks that each layer may encounter [9]. A crucial part of IoT security is played by the sensor layer, which is at the forefront of data collecting.

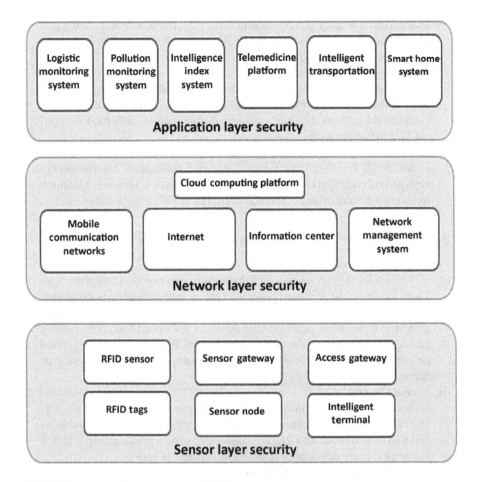

FIGURE 9.1   Security framework of IoT.

9.3.2 SENSOR LAYER'S SECURITY ISSUES

Compared to traditional network sensor nodes, sensor networks in the IoT implemented in an unsupervised environment include a few novel features.

i. **Weak Signal Strength**

Most sensor nodes operate in low power, long-time conditions, and communicate with one another primarily using wireless networks [11]. The signal during wireless communication is often easily impacted by disruptive waves [12,13]. Therefore, using a wireless network to send information is not secure.

ii. **Exposed Node**

Wireless channels are open and shared, which presents challenges for exposed workstations and hidden terminals [14]. For instance, when RFID technology is employed in the sensor layer [15], the sensor node is the perfect location for attackers since the RFID-enabled gadget will be hidden not just by its owner but also by others.

iii. **Dynamic Network Topology**

IoT node positions regularly shift from one location to another. All network monitoring and cyber security solutions must compact with additional complicated network data and stricter real-time requirements than the typical TCP/IP network [16].

iv. **Limited Computing and Storage Capacity**

IoT nodes typically have minimal power consumption, limited energy storage and computational capabilities [17,18]. Thus, a seamless transition of security systems from traditional networks to IoT is not possible.

9.3.3 DIFFERENT SECURITY MECHANISMS FOR SENSOR LAYER

i. **Encryption Mechanism**

The underlying principle of information security is a cryptosystem. There are two most advanced types of cryptography applications in conventional networks, both end-to-end and point-to-point encryption. According to the IoT framework, , a single-chip device or low-speed central processing unit (CPU) serves as the sensor layer node [19]. Programs that encrypt and decode data do not require a lot of power or storage. Therefore, IoT encryption mechanisms should be simple.

ii. **Access Control**

Access control mechanisms have gained some new meanings in the IoT. In an IoT network, a "machine" should be permitted to access the system as opposed to a "human" in a TCP/IP network [20]. As a result, it must allocate and send sharing data that is self-determined across nodes.

iii. **Authentication Mechanism of Nodes**

The receiver uses an authentication technique to confirm the sender's actual identity and check for data modification during transmission. A sensor layer authentication technique is required from the perspective of IoT architecture [21]. Data encryption can prevent intruders from thieving and altering sensitive data via keeping the information secure by encoding it [14]. Authentication can confirm that the genuine node is functioning.

9.4 DYNAMIC CIPHER

A "Dynamic cipher security certificate" is an authentication method with variable key security that is rely on a request-reply system. Table 9.1 illustrates its fundamentals, and all communication participants use the same key matrix.

The key matrix's storage area is 8 * 8 * 8 = 256 Bytes. A coordinate with a length of 1–16 bits will be generated at random by the parties talking. The length of the random password that results from this coordinate will range from 4 to 256 bytes, and therefore, theoretically, there are 64! = 1.26 * 1,089 passwords. It genuinely operates as a "one-time one cipher." The key itself is not sent; rather, the communication

TABLE 9.1

Key Matrix

	1	2	3	4	5	6	7	8
A	34gg	fve5	54f4	4c57	5658	Ad65	58gd	4534
B	4atf	4a4a	45dh	45ac	5e5g	54de	vw5s	5egw
C	hsfs	dv6d	345d	ef2f	45bt	4eds	4de3	4fgw
D	k4fs	s3g4	56gg	da45	vfgw	54gg	4dg4	afa3
E	jf3t	3afv	3fad	3dfv	5xag	45sa	hfy4	ddff
F	Ujk4	4aav	4geg	4fd4	dq53	45sw	gwtr	4a45
G	32gl	4a45	3fed	cts6	Sg55	dg5g	36fd	gf34
H	o36g	ddfg	sfey	ef4t	ca45	gfd5	gsdh	w3gr

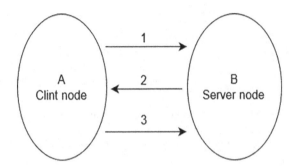

FIGURE 9.2   Three-step authentication.

parties merely exchange the key coordinate. All of its work uses a "one-time one-cipher" that dynamically assembles arbitrary coordinates and a key matrix to secure communication between parties.

9.4.1 AUTHENTICATION PROCESS

Figure 9.2 depicts the Dynamic cipher security certificate's authentication procedure. A and B are two communication nodes such as server and client node.

The procedure is as follows:

1. A→B: $Pos_{x1,y1}$, E (K_{ab1}: ID_a, Cmd, Ta1)
2. B→A: $Posx_{2,y2}$, E (K_{ab2}: ID_b, $C_{m,x,y}$, Ta1,Tb1)
3. A→B: E (Kab3: Text, ID_A, Ta2, Tb1)

Where Cmd denotes a connection request, Pos x y denotes the key matrix's location, Ta and Tb denote the nodes A and B's timestamps, and ID a and ID b denote the nodes A and B's respective ID numbers. E (kab:m) denotes encoding a message using the password kab, and Text denotes a constant message. As we can see from the description above, client A initiates a timer to wait for a response from server B while sending an encrypted ID a, a connection request, and timestamp Ta1. A will end this session if there is no echo response. When server B receives some data, it will check the ID from node A. If A is legitimate, then server B sends encrypted data to A, including ID b, the coordinates of the key matrix, timestamps Ta1 and Tb1, at the same time as client B starts a countdown in anticipation of client A's response. B will end this session if no echo reply is received. When client A receives some information, it will check the Ta1 coming from node B. A receives a communication password, Kab, based on the coordinates of the key matrix, if B is genuine. Then A creates a new timestamp Ta2 and combines it with Tb1 before transmitting the message and all this data will be sent to server B. We have created a communication channel between the parties up to this point. One-time one-ciphers or fixed passwords should be used while sending data.

9.4.2 ENCRYPTION MECHANISMS

Typically, we employ a by-hop encryption strategy on the network layer, which encrypts data during transmission but requires each node to maintain plaintext throughout decryption and encryption. In the conventional application layer encryption system, data is only explicitly disclosed to the sender and the receiver, and it is always encrypted throughout transmission and at forwarding nodes.

Due to the close relationship between the application and network layers in the IoT architecture, we must decide between by-hop and end-to-end encryption. By-hop encryption may be used at the network layer to implement various applications safely, and thus, if we utilize it, we can only encrypt the links that need to be confined [22]. The ease for end users comes from the security mechanism being transparent to business applications in this way. The benefits of the by-hop, such as reduced latency, good effectiveness, low price, and so on, are now well appreciated. "By-hop encryption" requires strong reliability of the communication nodes because each node can obtain the plaintext message due to the decryption process at the transmission node. End-to-end encryption enables us to select a different security policy based on the kind of company, meeting the high-security needs of the latter with a high level of protection. The destination address cannot be encrypted because each node chooses how to distribute messages based on the destination address, which prevents end-to-end encryption from doing so. As a result, hostile assaults are unable to hide the message's origin and final destination. From the aforementioned research, it can be inferred that end-to-end encryption is the best option when the security demands of a business are high. When the security needs of a business are low, by-hop encryption protection can be used. Therefore, a different encryption scheme based on the various criteria is used. There is still a lot of work to be done in this field of research since the IoT is still in its early phases of development and the study of safety measures is insufficient.

9.4.3 CRYPTOGRAPHIC ALGORITHMS

A renowned and well-respected set of cryptographic algorithms has been used in Internet security protocols which is demonstrated in Table 9.2.

Data is typically encrypted for confidentiality using the "Advanced Encryption Standard (AES) block cypher" [23]; key agreements are obtained using the "Diffie-Hellman" asymmetric key agreement method; digital signatures and transport keys are frequently established using the Rivest Shamir Adelman (RSA) asymmetric algorithm; and integrality is determined using the Secure Hash method using SHA-1 and SHA-256. The elliptic curve cryptography (ECC) technique is another important asymmetric method [2]. ECC can give equal safety by using a lower-length key, but its use has stalled recently. A CPU with enough speed and memory must be available to implement these cryptographic methods. Because it is unclear how to implement these cryptographic approaches to the IoT, more study is required to verify that algorithms can be effectively implemented in IoT devices with limited memory and slow CPUs.

Here, Table 9.3 illustrates the comparison of various cryptographic algorithms, feedback, and their applications.

TABLE 9.2

A Suit of Cryptographic Algorithms

Algorithm	Purpose
Advanced Encryption Scheme	Confidentiality
Rivest Shamir Adelman (RSA)/Elliptic Curve Cryptography (ECC)	Digital signature key transport
Diffie-Hellman (DH)	Key agreement
Secure Hash Algorithm SHA-1/SHA-256	Integrality

TABLE 9.3

Comparisons of Various Cryptographic Algorithms [23]

Algorithm	Compared with	Feedback	Application
ECC	Symmetric cryptographic algorithms	The idea of co-designing hardware and software is raised; asymmetry calls for more powerful CPUs. Nonetheless, does well	Pervasive computing
AES	Hardware findings compared to previously obtained results	AES-128 has been effectively applied on three distinct platforms; however, using a T-table on a GPU is a different strategy	RFID
PRESENT	AES, SEA, ICEBERG	Different architecture: Round-based data path is more appropriate for RFID and new technologies are used as compared to Pipelined and Minimal data paths	RFID

(*Continued*)

TABLE 9.3 (Continued)
Comparisons of Various Cryptographic Algorithms [23]

Algorithm	Compared with	Feedback	Application
HUMMIG-BIRD	PRESENT	On this platform, an active and passive RFID tag security solution that is superior to PRESENT is provided	RFID
PHOTON	SPONGENT, KECCAK-200, KECCAK-400	A significant impact of round functions on algorithms. PHOTON is the worst in terms of throughput/area, whereas SPONGENT is the best. However, PHOTON can scale sustainably	RFID tags
DESL	AES-128, HIGHT	Robust to numerous DES assaults that are vulnerable, better for RFID tags, and with low GE counts	RFID
HIGHT	FPGA Scalar and FPGA pipeline architecture	Two styles high: Implementation and comparison of scalar and pipeline designs	RFID using FPGA
TEA	SEA, GEA, EA, RR	GEA uses the least amount of overall energy, while TEA is the second-best server in terms of energy efficiency	Energy-conscious server-picking methods in a scalable cluster
LEA	PRESENT, Hummingbird, Ktantan, DESL, AES, LED	Although not the greatest (but in a higher place) among throughput/area, the Speed opt version is quite effective; yet, it excels in throughput output alone	RFID using FPGA's
Simon	AES, PRESENT, SPECK, TWINE, PRINCE	Simon and Speck are superior to AES in terms of implementation and are suited for usage with diverse networks. Very productive to work with it	ASIC application
SPECK	Speck and Simon 32, 48, 64, 96, 128	SCADA system implemented on a PLC. There are two forms of data: DWORD and BYTE 3	Not available
TWINE	AES, PRESENT, HIGHT, Piccolo	The key schedule in TWINE-80/128 is used by saturation cryptanalysis and impossible differential cryptanalysis	Not available

9.5   CONCLUSION AND FUTURE WORK

In this chapter, a dynamic cipher security certificate has been proposed along with its application. This protocol is implemented on a key-matrix-based "one-time one-cipher" way of communication. Its encryption and decryption procedures both take little space. Time stamping technology may be used by parties that are communicating to ensure their real-time. As already indicated the dynamic variable cipher security certificate may have a stronger use case in the IoT's sensor layer. The following stages in creating a cryptographic cipher for the IoT are to either improve the ones that already exist or create a new model based on the suggested strategy while using a cutting-edge encryption standard such as AES.

REFERENCES

[1] A. Botta, W. De Donato, V. Persico, and A. Pescapé, "Integration of cloud computing and Internet of Things: A survey," *Futur. Gener. Comput. Syst.*, vol. 56, pp. 684–700, 2016, doi:10.1016/j.future.2015.09.021.

[2] L. Marin, M. P. Pawlowski, and A. Jara, "Optimized ECC implementation for secure communication between heterogeneous IoT devices," *Sensors (Switzerland)*, vol. 15, no. 9, pp. 21478–21499, 2015, doi:10.3390/s150921478.

[3] X. Jia, Q. Feng, T. Fan, and Q. Lei, "RFID technology and its applications in Internet of Things (IoT)," *2012 2nd Int. Conf. Consum. Electron. Commun. Networks, CECNet 2012- Proc.*, Yichang, China. pp. 1282–1285, 2012, doi:10.1109/CECNet.2012.6201508.

[4] S. Sicari, A. Rizzardi, L. A. Grieco, and A. Coen-Porisini, "Security, privacy and trust in Internet of things: The road ahead," *Comput. Networks*, vol. 76, no. November, pp. 146–164, 2015, doi:10.1016/j.comnet.2014.11.008.

[5] P. Cheema and N. Julka, "Dynamic cipher for enhanced cryptography and communication for Internet of Things," *Lect. Notes Comput. Sci. (including Subser. Lect. Notes Artif. Intell. Lect. Notes Bioinformatics)*, vol. 10618 LNCS, no. 8, pp. 84–94, 2017, doi: 10.1007/978-3-319-69155-8_6.

[6] M. Al-Zubaidie, "Implication of lightweight and robust hash function to support key exchange in health sensor networks," *Symmetry (Basel)*, vol. 15, no. 1, 152. 2023, doi:10.3390/sym15010152.

[7] J. Park, R. Sandhu, and L. Fellow, "Activity control design principles: Next generation access control for smart and collaborative systems," *IEEE Access*, vol. 9, pp. 151004–151022, 2021, doi:10.1109/ACCESS.2021.3126201.

[8] J. J. Hathaliya and S. Tanwar, "An exhaustive survey on security and privacy issues in Healthcare 4.0," *Comput. Commun.*, vol. 153, no. February, pp. 311–335, 2020, doi:10.1016/j.comcom.2020.02.018.

[9] A. Whitmore, A. Agarwal, and L. Da Xu, "The Internet of Things-A survey of topics and trends," *Inf. Syst. Front.*, vol. 17, no. 2, pp. 261–274, 2015, doi:10.1007/s10796-014-9489-2.

[10] M. Al-Zubaidie, Z. Zhang, and J. Zhang, "REISCH: Incorporating lightweight and reliable algorithms into healthcare applications of wsns," *Appl. Sci.*, vol. 10, no. 6, 2020, doi:10.3390/app10062007.

[11] H. Shafagh, "Toward computing over encrypted data in IoT systems," *XRDS Crossroads, ACM Mag. Students*, vol. 22, no. 2, pp. 48–52, 2015, doi:10.1145/2845157.

[12] J. M. Bohli, R. Kurpatov, and M. Schmidt, "Selective decryption of outsourced IoT data," *IEEE World Forum Internet Things, WF-IoT 2015- Proc.*, Milan, Italy. pp. 739–744, 2015, doi:10.1109/WF-IoT.2015.7389146.

[13] D. Altolini, V. Lakkundi, N. Bui, C. Tapparello, and M. Rossi, "Low power link layer security for IoT: Implementation and performance analysis," *2013 9th Int. Wirel. Commun. Mob. Comput. Conf. IWCMC 2013*, September 2014, pp. 919–925, 2013, doi:10.1109/IWCMC.2013.6583680.

[14] J. Granjal, E. Monteiro, and J. Sa Silva, "Security for the internet of things: A survey of existing protocols and open research issues," *IEEE Commun. Surv. Tutorials*, vol. 17, no. 3, pp. 1294–1312, 2015, doi:10.1109/COMST.2015.2388550.

[15] Q. Jing, A. V. Vasilakos, J. Wan, J. Lu, and D. Qiu, "Security of the Internet of Things: Perspectives and challenges," *Wirel. Networks*, vol. 20, no. 8, pp. 2481–2501, 2014, doi:10.1007/s11276-014-0761-7.

[16] H. Shafagh, A. Hithnawi, A. Dröscher, S. Duquennoy, and W. Hu, "Talos: Encrypted query processing for the Internet of Things," *SenSys 2015- Proc. 13th ACM Conf. Embed. Networked Sens. Syst.*, Seoul, South Korea, pp. 197–210, 2015, doi:10.1145/2809695.2809723.

[17] N. Verma, S. Singh, and D. Prasad, "A review on existing IoT architecture and communication protocols used in healthcare monitoring system," *J. Inst. Eng. Ser. B*, 103, 245–257, 2021, doi:10.1007/s40031-021-00632-3.

[18] W. Shi, N. Kumar, P. Gong, N. Chilamkurti, and H. Chang, "On the security of a certificateless online/offline signcryption for Internet of Things," *Peer-to-Peer Netw. Appl.*, vol. 8, no. 5, pp. 881–885, 2015, doi:10.1007/s12083-014-0249-3.

[19] A. Patil, G. Bansod, and N. Pisharoty, "Hybrid lightweight and robust encryption design for security in IoT," *Int. J. Secur. Appl.*, vol. 9, no. 12, pp. 85–98, 2015, doi:10.14257/ijsia.2015.9.12.10.

[20] H. J. Kim and K. Kim, "Preliminary design of a novel lightweight authenticated encryption scheme based on the sponge function," Proc. -2015 10th Asia Jt. Conf. Inf. Secur. AsiaJCIS 2015, Kaohsiung City, Taiwan, pp. 110–111, 2015, doi:10.1109/AsiaJCIS.2015.24.

[21] N. Verma, S. Sangwan, S. Sangwan, and D. Parsad, "IoT security challenges and counters measures," *Int. J. Recent Technol. Eng.*, vol. 8, no. 3, pp. 1519–1528, 2019, doi:10.35940/ijrte.C4212.098319.

[22] I. I. C Toma, C. Ciurea, "Authentication issues for sensors in IoT solutions," *Proc. 6th Int. Conf. Secur. Inf. Technol. Commun.*, Bucharest, Romania. pp. 2285–1798, 2013.

[23] C. C. Aggarwal, *Managing & Mining Sensor Data*, Springer-Verlag New York Inc, Midtown Manhattan, New York City, 2014. doi:10.1007/978-1-4614-6309-2.

10 Smart Cities Unveiled
The AI Revolution in Urban Development

Pratibha, Gaganpreet Kaur,
Padam Dev, and Amandeep Kaur

10.1 INTRODUCTION

The 21st century has seen an unprecedented global shift towards urbanisation, with more than half of the world's population now residing in cities. This transformation has necessitated innovative approaches to urban planning and development, giving rise to the concept of "Smart Cities." Smart cities are urban environments that leverage advanced technologies to enhance the quality of life for their residents while addressing the challenges of modern urban living [1].

The utilisation of artificial intelligence (AI) within smart cities and its impact on governance, decision-making processes, creative practices, and potential for transformative change have been the subject of extensive discourse and practical exploration in recent years. The utilisation of AI enables the generation of data in both government and private sectors, facilitating the exploration of novel methodologies to enhance our comprehension of the world. The utilisation of big data has the potential to enhance resource allocation and facilitate informed decision-making; integration of AI with the Internet of Things has the potential to exert a favourable impact on the process of smart decision-making. Currently, AI is increasingly essential for both everyday life and organisational processes. This is due to significant technological advancements that have greatly enhanced the development and capabilities. AI plays a significant role in enhancing the decision-making processes of smart cities. This is achieved by the use of systematic and organised approaches to data collection, as well as the utilisation of rational decision-making systems. Unlike relying on intuition, trial and error, or generalising from past experiences, smart decision-making leverages AI to ensure more accurate and informed decisions [2].

10.1.1 PILLARS OF A SMART CITY

The term "smart city domain" pertains to a distinct area of concentration within the wider framework of smart city advancement. Smart cities leverage advanced technologies and data-driven approaches to augment the quality of urban living. These innovations are implemented across several domains, representing distinct sectors within the urban landscape (Figure 10.1).

DOI: 10.1201/9781032673479-13

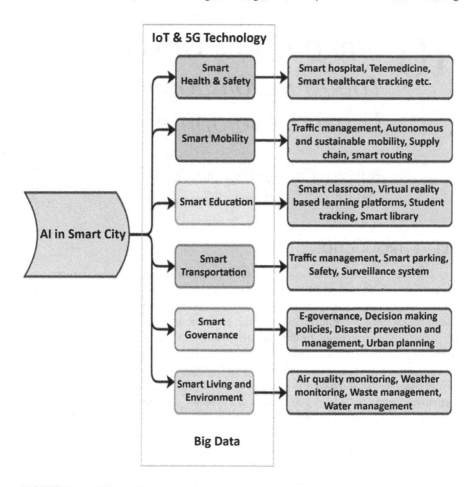

FIGURE 10.1   Pillars of smart city.

Within these specific categories, urban areas strive to enhance effectiveness, sustainability, and the holistic well-being of their inhabitants. The domains commonly associated with smart cities encompass various areas such as transportation, energy management, public safety, waste management, smart living and environment, infrastructure maintenance, public services and governance, urban planning, public health, water management, and education.

Each domain signifies a distinct aspect of urban existence wherein technology, data analysis, and AI are utilised to foster interconnectedness, enhance efficiency, and promote responsiveness within cities. These several fields all contribute to the overarching concept of smart cities, wherein technology and data play crucial roles in driving growth and fostering innovation [3].

10.1.2 ARTIFICIAL INTELLIGENCE IN SMART CITIES

Smart cities utilise a range of AI methodologies to optimise operational effectiveness, promote sustainability, and improve the overall well-being of inhabitants. AI systems are capable of assimilating data obtained from Internet of Things (IoT) devices and sensors in order to effectively oversee and regulate multiple facets of urban environments, including but not limited to air quality, traffic patterns, and energy usage [1,4].

Several commonly employed AI techniques in the context of smart cities encompass:

Machine Learning (ML) refers to the utilisation of algorithms in order to analyse data derived from sensors, cameras, and other sources. This analytical process enables the generation of predictions and the optimisation of services in domains such as traffic management, energy consumption, and public safety. Deep learning, which is a subfield of ML, is used in various domains, including security camera image recognition and natural language processing for chatbots and virtual assistants [5,6].

Natural language processing is a field of study that empowers AI systems to comprehend and generate responses to human language. This capability facilitates seamless interactions between AI systems, such as chatbots and voice assistants, and individuals. Computer Vision: Computer vision techniques are employed to analyse images and videos in applications such as surveillance, traffic management, and environmental monitoring [7]. Reinforcement Learning: In transportation and autonomous vehicles, reinforcement learning helps systems learn optimal routes and behaviours over time [6].

AI systems effectively amalgamate data derived from IoT devices and sensors to actively oversee and regulate diverse facets of urban environments, including but not limited to air quality, traffic patterns, and energy usage (Figure 10.2).

Pattern recognition is a field that utilises AI techniques to find abnormalities or trends within data. This process is particularly useful in several domains such as predictive maintenance, security, and public safety. Optimisation algorithms are employed in the optimisation of traffic signal sequences, public transportation routes, and garbage collection schedules. Data mining is a computational technique employed to extract significant insights from extensive databases, with the purpose of informing decision-making processes in several domains, including urban planning and healthcare [7,12].

Edge computing refers to the practise of conducting AI processing in close proximity to the data source, commonly referred to as the "edge." This approach aims to minimise latency and enhance the ability to make real-time decisions, which holds significant importance in many applications such as autonomous vehicles and emergency response systems [8,9].

The field of Robotics and Autonomous Systems involves the utilisation of AI to operate robots and autonomous systems in many applications, such as garbage collection, surveillance, and infrastructure maintenance [8]. Smart grids utilise AI to

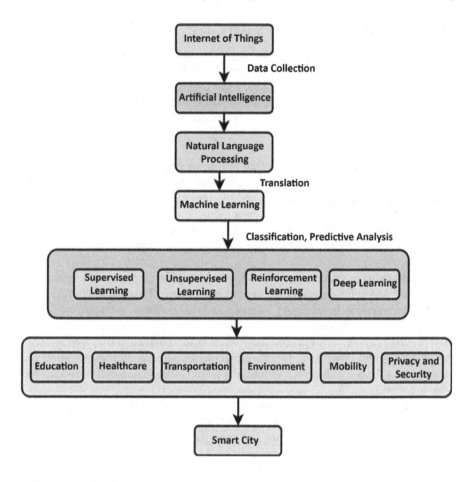

FIGURE 10.2 Conceptual structure of smart city with AI.

effectively manage and enhance the distribution of energy, resulting in enhanced energy efficiency and reliability over a period of time [11].

Predictive analytics involves the utilisation of AI models to generate predictions by analysing both historical and real-time data. These forecasts serve to facilitate various processes such as planning, decision-making, and resource allocation.

10.1.3 IoT AND AI IN SMART CITY

The integration of IoT and AI technologies in the manufacturing sector has emerged as a significant area of research and development [2]. The integration of IoT and AI in the industrial industry, known as "Smart Manufacturing" or "Industry 4.0," has revolutionised product design, production, and monitoring processes, enhancing operational effectiveness and product standards oversight. IoT in manufacturing industry. The implementation of IoT technology in the industrial sector involves strategically placing sensors and interconnected devices throughout the production process to gather data on equipment performance, ambient conditions, and product quality, which is then communicated to a centralised system.

- **Predictive Maintenance:** IoT sensors enable real-time monitoring of machinery, enabling predictive maintenance. AI systems analyse data, predicting maintenance timing, reducing operational inactivity and preventing costly equipment failures [9].
- **Quality Control:** The use of IoT-enabled sensors and AI algorithms in the production process ensures consistent product quality checks at various stages, allowing for real-time modifications to address flaws and deviations from established standards [7].
- **Inventory Management**: IoT technology in inventory management systems allows real-time inventory monitoring, ensuring timely availability of raw materials and components. AI algorithms optimise inventory levels and reduce wastage [8].
- **Energy Efficiency:** IoT sensors improve manufacturing efficiency by monitoring energy consumption, while AI optimises consumption to reduce expenses and minimise ecological footprint.
- AI enhances the IoT in the manufacturing sector by offering the cognitive powers and decision-making power required to effectively use the gathered data. One of the primary areas where AI is extensively utilised is in the domain of manufacturing [10,11].
- **Process Optimisation:** It involves the utilisation of AI to analyse data from the IoT in order to discover inefficiencies within a given process and provide recommendations for improvements. Consequently, this phenomenon results in heightened levels of production and enhanced utilisation of resources [10].
- **Predictive Analytics:** AI models use IoT data to predict demand, quality issues, and equipment breakdowns, enabling manufacturers to take proactive measures and minimise waste generation and operational downtime, thereby enhancing efficiency and productivity.
- **Robotics and Automation:** AI-powered robots and autonomous systems are increasingly used in manufacturing to perform intricate operations accurately, improving product uniformity and reducing human labour costs [8].
- **Supply Chain Optimisation:** AI can optimise supply chains by analysing IoT data, identifying areas for improvement, reducing lead times, and enhancing overall efficiency [2].

    The integration of IoT and AI in manufacturing presents advantages but also presents challenges. Data security is crucial for safeguarding industrial information. Enhancing workforce skills is essential for AI and IoT technologies. Integration complexity presents financial implications, necessitating standardised protocols and best practices. Privacy concerns arise, especially regarding worker data collection [12].

10.1.4 Smart Healthcare

The utilisation of AI technology in the healthcare sector has led to a significant transformation of the medical area. This technology facilitates disease diagnosis through the analysis of medical images, expedites the process of drug discovery, allows for remote monitoring of patients, and provides personalised recommendations for therapy. AI is capable of forecasting illness outbreaks and enhancing the efficiency of hospital operations. The utilisation of AI-driven chatbots facilitates the optimisation of medical records administration and promotes increased patient involvement. Moreover, robotic surgical devices have the capability to augment surgical precision. The responsible utilisation of AI plays a crucial role in effectively addressing ethical and privacy concerns that arise in healthcare applications [17].

10.1.5 Smart Education

The advent of AI has brought about a paradigm shift in the field of education, resulting in a fundamental restructuring of the methods and systems used for delivering and overseeing learning experiences. The education sector has greatly benefited from the significant contributions of AI, which include the provision of personalised learning experiences, adaptive examinations, and automated grading systems. Furthermore, the utilisation of AI-powered virtual tutors and chatbots provides students with uninterrupted assistance, so ensuring that the process of learning is not restricted by conventional timetables [16].

AI also utilises data analytics to offer significant insights, empowering institutions to make informed decisions based on data for the purpose of enhancing curriculum design and educational outcomes. The utilisation of technology enhances inclusivity by increasing the accessibility of educational resources to a wide range of learners, including individuals with impairments. In addition, AI has enhanced the efficiency of content generation and professional development in the field of education. Through the creation of educational resources and the implementation of customised training, AI enhances the efficacy of teachers within the classroom setting.

The integration of various AI technologies in this context seeks to enhance the efficiency and effectiveness of both the process of acquiring knowledge and the act of imparting knowledge. However, it is vital to consider the ethical and privacy components to enable fair access and responsible usage of AI in education [19].

10.1.6 Smart Transportation

Smart transportation encompasses the utilisation of sophisticated technologies to augment the effectiveness, security, and ecological viability of transportation networks. This methodology utilises digital advancements to establish transportation networks that are more interconnected and intelligent. Smart transportation systems leverage real-time data, IoT devices, and AI to enhance traffic management, mitigate congestion, minimise emissions, and enhance overall mobility. The fundamental elements of intelligent transportation encompass intelligent traffic control systems, Global Positioning System navigation technology, autonomous cars, and real-time transit data. The objective is to develop a transportation experience that is characterised by enhanced integration, ecological sustainability, and a focus on user needs, hence yielding advantages for both individuals and communities [14,17].

10.1.7 SMART ENVIRONMENT

The principal objectives of intelligent settings are to augment convenience, mitigate energy usage, minimise environmental repercussions, and foster well-being. They play a significant role in the development of habitable and environmentally sustainable environments that are responsive to the requirements of both the inhabitants and the surrounding ecological system [17].

10.1.8 MOBILITY

AI plays a crucial role in the advancement of smart mobility through its ability to optimise traffic flow, accurately predict congestion, and boost the efficiency of public transportation systems. This technology facilitates the implementation of ride-sharing services, autonomous vehicles, and effective parking strategies. The utilisation of AI in the development of user-centric features provides individuals with tailored travel advice, timely updates, and environmentally conscious alternatives. Additionally, it facilitates data-driven decision-making and improves safety and security inside transportation networks. areas. Smart mobility refers to the utilisation of advanced technologies and data analysis to tackle the various issues associated with urban transportation, including but not limited to traffic congestion, air pollution, and limited mobility choices. Its primary objective is to improve the overall quality of life in urban areas [13,17].

10.1.9 PRIVACY AND SECURITY

Privacy is protected by means of data anonymisation, which permits the study of data without disclosing any personal information. Resident privacy controls provide individuals with the ability to exert influence over the use of their personal information [17].

AI functions as a vigilant protector, consistently observing and analysing several systems in order to identify and address potential dangers. The utilisation of real-time data from many sources enhances emergency response capabilities, facilitating prompt reactions to situations. The seamless incorporation of AI within smart cities plays a pivotal role in enhancing the safety of urban areas while simultaneously upholding the fundamental rights to privacy of individuals [18].

10.2   RISE OF AI

Figure 10.3 illustrates the temporal evolution of the search terms 'Smart City' and 'Artificial Intelligence' as observed in Google Trends from 2018 onwards. The above

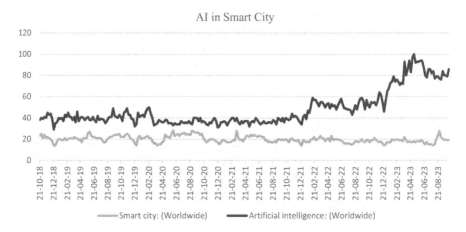

FIGURE 10.3   Smart city and artificial intelligence from 2018, as sourced from Google Trend.

picture depicts a notable rise in the level of interest surrounding AI and smart cities as observed over a period of time

The development of a smart city is intricately connected to the widespread and efficient utilisation of the Internet. The use of technologies and data-driven solutions facilitates the integration of urban living enhancements, sustainability promotion, and streamlining of diverse city processes, ultimately leading to an enhancement in the overall well-being of the citizens.

Figure 10.4 shows the number of global Internet users has experienced a progressive increase over the past few decades. According to the International Telecommunication Union (ITU), it is projected that almost 5.4 billion individuals, or for around 67% of the global population, would be utilising the Internet by the year 2023. The aforementioned figure demonstrates a growth of 45% from the year 2018, with an estimated 1.7 billion individuals having gained Internet access within that timeframe [15].

The International Institute for Management Development (IMD) Smart City Index (SCI) is commonly employed to evaluate and classify cities according to a range of criteria pertaining to their technological and socioeconomic progress, with the aim of ascertaining their level of smartness. Table 10.1 shows the ranks of the top 20 cities as per the Smart City Index for the years 2019–2023. These rankings have been determined using the most up-to-date methodology available (Figure 10.5).

Notably, it is observed that among the top 20 cities, a significant majority of 17 cities have been affiliated with the SCI since its inception. Out of the total of 17 cities examined, 6 of them demonstrate either a consistent upward trend or a state of stability when comparing their performance on a year-to-year basis. The cities identified as "super-champions" are Zurich, Oslo, Singapore, Beijing, Seoul, and Hong Kong [16].

Individuals using the Internet

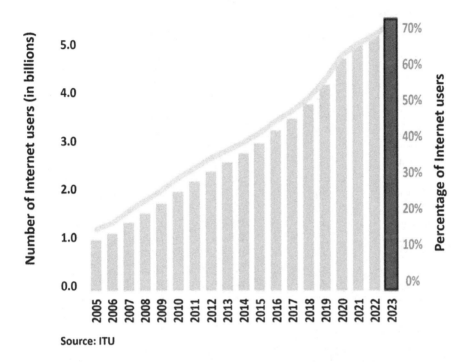

Source: ITU

FIGURE 10.4 Internet usage [15].

10.3 CHALLENGES AND ISSUES IN SMART CITY WITH AI

The integration of AI in the creation of smart cities presents a range of challenges and issues that necessitate attention and resolution (Figure 10.6).

- The susceptibility of smart cities to cyberattacks renders them highly appealing targets for unauthorised access to data. Ensuring the security of data and systems that are dependent on AI is of paramount importance in order to mitigate the risks associated with data breaches and potential interruptions to essential services [10].
- **Privacy Concerns**: The extensive utilisation of AI inside smart cities results in the generation of substantial volumes of data, hence giving rise to apprehensions over the privacy of inhabitants. Ensuring the presence of comprehensive data protection and privacy rules is vital in order to effectively preserve individuals' personal information [11].

TABLE 10.1

Top 20 cities as per the SCI Smart City Index for the years 2019–2023 [16]

City	Rank 2023	Rank 2021	Rank 2020	Rank 2019
Zurich	1	1	1	1
Oslo	2	2	2	2
Canberra	3	—	—	—
Copenhagen	4	5	3	4
Lausanne	5	4	—	—
London	6	3	10	3
Singapore	7	7	7	10
Helsinki	8	9	5	6
Geneva	9	6	8	7
Stockholm	10	11	9	9
Hamburg	11	8	6	—
Beijing	12	17	22	30
Abu Dhabi	13	12	14	16
Prague	14	10	4	8
Amsterdam	15	13	11	11
Seoul	16	18	20	23
Dubai	17	14	19	13
Sydney	18	29	32	22
Hong Kong	19	33	34	38

- Digital inclusion refers to the equitable access and utilisation of AI-driven services across inhabitants, highlighting the potential for these technologies to amplify existing socioeconomic disparities. It is imperative to exert endeavours in order to guarantee equitable distribution of AI advantages across all segments of the community, irrespective of their socioeconomic standing [12].
- The ethical utilisation of AI presents a considerable obstacle in ensuring equitable treatment and the prevention of discrimination against specific demographics. The presence of bias inside AI algorithms has the potential to sustain and reinforce existing societal inequities, hence resulting in unjust treatment [13].
- Interoperability is a common characteristic observed in smart cities, wherein a diverse range of technologies and systems are frequently integrated. The smooth integration of these technologies poses a substantial difficulty. The establishment of interoperability standards is necessary in order to facilitate the process of integration [14].
- The cost associated with the implementation of AI systems can be substantial, encompassing both the initial development and ongoing maintenance expenses. Urban areas may have challenges in obtaining the requisite

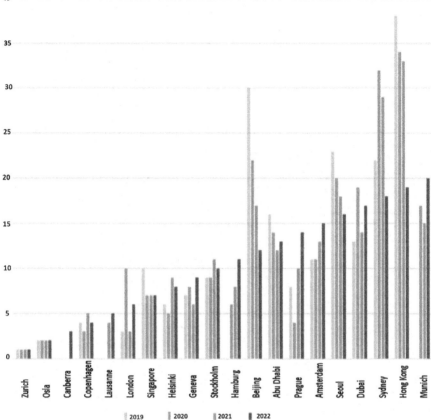

FIGURE 10.5 Top 20 smart cities according smart city index 2023 with methodology applied [15].

financial resources and specialised knowledge for the implementation of AI initiatives. The identification of viable funding models for smart city programmes poses a significant problem [17].

- The efficacy of AI is contingent upon the calibre of the data upon which it is reliant. The presence of inaccurate or incomplete data has the potential to result in erroneous conclusions and judgements, hence affecting the overall performance of AI-driven services [18].
- AI systems are susceptible to security vulnerabilities and can be targeted by malicious attacks. Adversarial actors have the capacity to engage in the manipulation of AI algorithms or exploit their vulnerabilities, which might result in potentially detrimental outcomes [17].

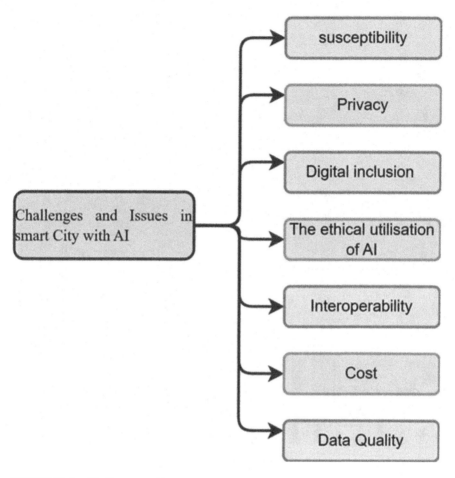

FIGURE 10.6 Challenges and issues in smart city with AI.

- The resolution of these challenges and issues necessitates a collective endeavour involving governmental bodies, industrial entities, and community groups. The establishment of explicit laws, rigorous standards, and a high degree of transparency in the utilisation of AI is imperative in order to foster the development of smart cities that are inclusive, secure, and efficient, thereby ensuring the well-being of all inhabitants [18].

10.4   FUTURE SCOPE

The urban issues faced by cities have reached unprecedented proportions due to the ever-increasing population and rapid population explosion in metropolitan areas. It is anticipated that there will be a continuation of the aforementioned trends, including further increases in pollution levels, resource shortages, traffic congestion, and other related issues. Contemporary urban areas are currently confronted with novel economic, political, and technological obligations that necessitate their fulfilment in order to ensure the provision of enduring prosperity to their inhabitants. It is imperative to capitalise on technological advancements and implement intelligent systems that can effectively maximise the utilisation of scarce resources.

The integration of AI in the development of a smart society represents a remarkable and transformative advancement towards the forthcoming era. AI transcends its technological nature and serves as the driving force behind the development of intelligent, efficient, and interconnected systems that are significantly impacting various aspects of our everyday lives, economies, and societal structures. AI has demonstrated its ability to enhance productivity, improve safety, and provide tailored experiences in various sectors, such as smart cities, healthcare, energy, transportation, agriculture, and retail, among others. Nevertheless, it is crucial to acknowledge and tackle the various concerns and challenges that arise in relation to AI.

REFERENCES

1. Kamruzzaman, M. M. "New opportunities, challenges, and applications of edge-AI for connected healthcare in smart cities." In *2021 IEEE Globecom Workshops (GC Wkshps)*, pp. 1–6. IEEE, Madrid, Spain, 2021.
2. Sterbenz, J. P. G. "Smart city and IoT resilience, survivability, and disruption tolerance: Challenges, modelling, and a survey of research opportunities." In *2017 9th International Workshop on Resilient Networks Design and Modeling (RNDM)*, pp. 1–6. IEEE, Alghero, Italy, 2017.
3. Ahwini, B. P., R. M. Savithramma, and R. Sumathi. "Artificial intelligence in smart city applications: An overview." In *2022 6th International Conference on Intelligent Computing and Control Systems (ICICCS)*, Madurai, India, pp. 986–993, 2022, doi:10.1109/ICICCS53718.2022.9788152.
4. Singh, P. D., G. Dhiman, and R. Sharma. "Internet of things for sustaining a smart and secure healthcare system." *Sustainable Computing: Informatics and Systems* 33 (2022): 100622.
5. Kaur, G., A. Kaur, and M. Khurana. "A stem to stern sentiment analysis emotion detection." In *2022 10th International Conference on Reliability, Infocom Technologies and Optimization (Trends and Future Directions) (ICRITO)*, pp. 1–5. IEEE, Greater Noida, India, 2022.
6. Kaur, G., A. Kaur, and M. Khurana. "A review of opinion mining techniques." *ECS Transactions* 107, no. 1 (2022): 10125.
7. Verma, P. and M. Mahajan. "Hierarchical clustering approach for modeling of reusability of function oriented software components". *PSRC, Planetary Scientific Research Center Proceeding. ISEMS, Bangkok.* (2011).
8. Lilhore, U. K., M. Poongodi, A. Kaur, S. Simaiya, A. D. Algarni, H. Elmannai, V. Vijayakumar, G. B. Tunze, and M. Hamdi. "Hybrid model for detection of cervical cancer using causal analysis and machine learning techniques." *Computational and Mathematical Methods in Medicine* 2022 (2022), pp. 1–17.
9. Rathore, P. S., J. M. Chatterjee, A. Kumar, and R. Sujatha. "Energy-efficient cluster head selection through relay approach for WSN." *The Journal of Supercomputing* 77 (2021): 7649–7675.
10. Ramesh, T. R., U. K. Lilhore, M. Poongodi, S. Simaiya, A. Kaur, and M. Hamdi. "Predictive analysis of heart diseases with machine learning approaches." *Malaysian Journal of Computer Science* (2022): 1, 132–148.

11. Joshi, S., S. Saxena, and T. Godbole. "Developing smart cities: An integrated framework." *Procedia Computer Science* 93 (2016): 902–909.

12. Dhiman, P., and A. Kaur. "Significance of convolutional neural network in fake content detection: A systematic survey." In *International Conference on Emergent Converging Technologies and Biomedical Systems*, Singapore, pp. 305–316. Springer Nature Singapore, 2022.

13. Gondal, A. U., M. I. Sadiq, T. Ali, M. Irfan, A. Shaf, M. Aamir, M. Shoaib, A. Glowacz, R. Tadeusiewicz, and E. Kantoch. "Real time multipurpose smart waste classification model for efficient recycling in smart cities using multilayer convolutional neural network and perceptron." *Sensors* 21, no. 14 (2021): 4916.

14. Herath, H. M. K. K. M. B., and M. Mittal. "Adoption of artificial intelligence in smart cities: A comprehensive review." *International Journal of Information Management Data Insights* 2, no. 1 (2022): 100076.

15. www.itu.int/en/ITU-D/Statistics (itu.int).

16. www.imd.org/smart-city-observatory/home/

17. Ahmad, K. A. B., H. Khujamatov, N. Akhmedov, M. Y. Bajuri, M. N. Ahmad, and A. Ahmadian. "Emerging trends and evolutions for smart city healthcare systems." *Sustainable Cities and Society* 80 (2022): 103695.

18. Al-Turjman, F., H. Zahmatkesh, and R. Shahroze. "An overview of security and privacy in smart cities' IoT communications." *Transactions on Emerging Telecommunications Technologies* 33, no. 3 (2022): e3677.

19. Khurana, M., I. Sharma, and G. Kaur. "Exploring 5G for vehicular networks." In *2022 Seventh International Conference on Parallel, Distributed and Grid Computing (PDGC)*, pp. 446–450. IEEE, Solan, Himachal Pradesh, India, 2022.

11 A Backstepping Neural Network-Based Manufacture and Leader-Follower Formation Control of Multiple Autonomous Underwater Vehicles

Manju Rani, Naveen Kumar, and Anita Dahiya

11.1 INTRODUCTION: BACKGROUND

AUV performance is a crucial element that must be considered for AUV formation control. AUVs come in a wide variety of shapes because they are made by diverse manufacturers to meet distinct needs. AUVs can be categorized based on their body types, intended uses, manufacturers, or scales. Comparing the four classifications of AUVs, body shape categorization is the most applicable. The main factors are as follows: (i) the majority of AUVs have several application objectives and are capable of handling multiple tasks simultaneously, and (ii) body scales or manufacturer-based classification is less effective and lacks classification features.

AUVs have a number of important characteristics, including speed, endurance, and operational depth. According to the literature review, manufacturer websites are a good place to find out more information about these characteristics of different-shaped AUVs. More activity has been seen in sectors ranging from research to AUV manufacturing and operation as a result of the recent appearance of substantial applications in both the commercial and military domains. As a result, component suppliers are making major development efforts to modify their product line to better support AUVs. As a result, AUV capabilities have rapidly increased and are expected to do so for some time.

Many undersea applications frequently require the employment of multiple AUV including deep sea archaeology, ocean floor surveys, oceanographic mapping, underwater pipeline inspection, underwater surveillance, and geological sampling

(Sahoo et al., 2019; Yan et al., 2019; Xiang et al., 2015; Qu et al., 2018; Kumar and Rani, 2020; Peng and Wang, 2017; Mancilla et al., 2022; Huang et al., 2016; Li and Yan, 2016; Bejarbaneh et al., 2020; Xu et al., 2022; Elhaki and Shojaei, 2020). .

How to control the formation to ensure that it stays in the correct configuration while the designated tasks are being done is the fundamental problem with multi-AUV cooperation. Due to AUVs' complicated dynamics and communication limitations, formation control has been the subject of research (Yuan et al., 2014; Wang et al., 2012). Different methods have been documented in the literature, such as the leader-follower approach (Yun et al., 2010), behavior-based control (Kumar and Stover, 2000; Balch and Arkin, 1998), virtual structure approach, decentralized coordinated control, behavior-based control, and the mechanics of artificial potential (Tong et al., 2000; Ge et al., 2004). The essential issue with multi-AUV collaboration is how to govern the formation so that it remains in the proper configuration while the assigned duties are being completed. The leader-follower strategy, in particular, has been the subject of extensive investigation because of its simplicity, reliability, and ease of application. The leader-follower approach, in particular, has garnered a lot of focus due to its dependability, simplicity, and convenience of usage.

An extensive literature review was done to find out more about controlling AUVs in leader-follower formation. To keep the formation intact, the leader AUV talks with the follower AUVs. These states are then sent via acoustic modem. The AUV control system experiences communication latency as a result of the acoustic modem's low data rate. The aforementioned issues have been resolved by a range of control techniques, including sliding mode control, NN-based control, adaptive/robust control, conventional control methods, backstepping control, and decentralized control techniques by the following references: Shojaei (2015), Lu et al. (2021), Bechlioulis et al. (2019), Emrani et al. (2010), Wang et al. (2020), Hou and Cheah (2009), Breivik et al. (2008), Riahifard et al. (2019), Atta and Subudhi (2013). In light of the formation control problem that many AUVs face, this study provided modeling results and stability assessment for six-degree-of-freedom AUVs. Shojaei (2015) proposed a leader-follower formation tracking controller for autonomous low-torque, underactuated sea surface vehicles under environmental disturbances. Hou and Cheah (2009) expanded on the concept of area control and created a simple PD with a gravity correction controller. This chapter included stability analysis and simulation results for six-degree-of-freedom AUV. Breivik et al. (2008) proposed a leader-follower based formation control by combining the integrator backstepping technique with the cascade theory. Riahifard et al. (2019) developed strong adaptive techniques to address the issue of underactuated autonomous surface vehicles' (ASVs) control of the leader-follower formation. Atta and Subudhi (2013) developed decentralized formation control for multiple AUVs with input saturation. Every AUV in the intended technique received information from its surrounding AUVs in accordance with the recommended communication architecture, and certain AUVs were able to access the desired path. Using adaptive algorithms, Lu et al. (2021) studied the ASV performance, which ensured formation tracking problems and guaranteed that all error signals would be evenly finally contained. Bechlioulis et al. (2019) introduced a distance-based aggressive decentralized design technique for leader-follower formation control of various autonomous submerged vehicles. The proposed method reduced complexity while offering robustness against model uncertainty. Emrani et al. (2010) investigated the problem of leader-follower formation control for multiple autonomous underwater vehicles (AUVs) in spatial motion. The goal of the project was to create a leader robot that could control the followers and follow a predetermined route while maintaining the leader's three-dimensional form. Wang et al. (2020) employed a neural adaptive formation control technique based on sliding mode to address an underactuated AUV leader-follower formation control problem. The suggested technique does not rely on any prior knowledge of hydrodynamic damping effects; rather, it is dependent on the leader's location information. The majority of the techniques discussed above are predicated on resilient, model-based, and adaptable control algorithms that are founded on precise comprehension of uncertain systems. Creating the regression matrix and getting reliable data is a real problem with the uncertain systems dynamic model. Using traditional model-dependent control approaches to guide the system becomes more challenging due to complex dynamics. Determining the bounds for outside disturbances and uncertainties is also important in relation to the sliding mode control (SMC) regulations. While any of these methods is insufficient, using a hydrodynamic matrix in particular is insufficient since it does not offer a dependable means of obtaining model parameters. This is due to the fact that the control rules, which could have a detrimental effect on the system's performance, heavily rely on the dynamic model. Due to their ability to overcome the shortcomings of the model-dependent controllers, the intelligent methods have been shown to be a more useful tool for the approximation of smooth non-linear functions that are unknown (Lu et al., 2020; Shojaei, 2016; Wang et al., 2019, 2020; Peng et al., 2011, 2012).

Neural network-based control technologies have several advantages, major among them being their adaptability. Increasingly complicated systems, such as those with self-learning capabilities, parallel distributed architecture, and non-linear mapping, have been managed successfully with the use of these technologies. NN-based control techniques for ASVs that are independent of models with underactuation were studied by Peng et al. (2011). In order to create a reliable adaptive control system for underactuated ASV formation control, Peng et al. (2012) combined it with a neural network. A unique approach to control the scaling of adaptive neuronal development for ASVs based on bearings was developed by Lu et al. (2020). This study demonstrated that under the condition that the stretched structure had a negligibly rigid bearing, the suggested formation approach could successfully carry out the anticipated formation scaling maneuver of ASVs. Huang et al. (2020) used radial basis function control and adaptive consolidation with reinforcement learning to build a number of under-actuated AUVs. This chapter's simulations and experimental findings demonstrated how successful the suggested control strategy is. Wang et al. (2019) created a neural adaptive SMC system that took into account both model uncertainties and outside disturbances to address the numerous underactuated AUVs' leader-follower formation control issues. Shojaei (2016) used robust, adaptable, multi-layer NNs to tackle the issue of the creation of autonomous marine surface boats, and it was shown that the suggested control strategy offered strength against uncertaintics brought on by sea waves and flows. We report on an efficient leader-follower formation control using backstepping neural networks for numerous AUV with unknown hydrodynamic parameters. Using the backstepping approach has the main benefit of offering resilience against uncertainty. The AUVs under development are fully-actuated, in contrast to the majority of leader-follower formation control systems that are currently in use.

Even though the significant control methods mentioned earlier have been demonstrated to achieve notable control performance, this is, as far as the author is aware, the first attempt in the literature to completely reduce the gap between the model-free and model-dependent control systems for leader-follower formation control of AUV. The systems dynamics in real-world applications are generally poorly understood, so this information is useful to take into account when constructing controllers. In the process of developing control laws, neural networks have proven to be a useful approximation for non-linear dynamics. As such, the present work aims to (i) leverage the current body of knowledge about partial system dynamics and (ii) leverage the critical role that NNs play as a tool to ascertain the unknown value of the dynamic component. These highlight the key contributions: (i) By using a model-free approach in combination with the partially available system dynamics data, this work offers a unique control mechanism for leader-follower formation control of multiple AUVs. (ii) The controller has an adaptive compensator to lessen the effects of outside disturbances and reconstruction errors. (iii) The stability of the closed-loop system is evaluated using the Lyapunov theorem and Barbalat's lemma. It is demonstrated that creation faults not only stay within the expected values but also asymptotically converge to a small neighborhood of zero. (iv) MATLAB simulation outcomes are utilized to compare the suggested method's effectiveness and dependability with the existing techniques.

11.2 DYNAMICAL MODEL DESCRIPTION

Numerous studies have planned the robust AUV model. Due to the inclusion of unmodeled elements and hydrodynamic components in the dynamics, the unique model proves to be incredibly complicated and profoundly non-linear. Given this, the most crucial feature of the AUV is identified, and to show the viability of the recommended control technique, a 4-DOF reduced unique model is then generated. To obtain the kinematic form of the AUV frameworks, we typically consider the earth-fixed inertial edge and the body-fixed edge casing. Typically, we employ two vectors for this Φ_i and χ_i to describe the status of the ith underwater vehicle. $\Phi_i = [\Phi_{1i}, \Phi_{2i}]$ is the ith vehicle's position and heading vector in an arbitrary/earth-fixed frame (A) in order to ensure that $\Phi_{1i} = [x_i, y_i, z_i]^T$ and $\Phi_{2i} = [\varphi_i, \theta_i, \psi_i]^T$. Next, $\chi_i = [\chi_{1i}, \chi_{2i}]$ is the body-fixed frame's vector of angular and linear velocities (B) with $\chi_{1i} = [U_i, V_i, W_i]^T$ and $\chi_{2i} = [P_i, Q_i, R_i]^T$, respectively. The following expression shows the ith AUV's location in the global coordinate frame:

$$\dot{\Phi}_i = J_i(\Phi_i)\chi_i = \begin{bmatrix} J_{1i}(\Phi_i) & 0_{3\times3} \\ 0_{3\times3} & J_{2i}(\Phi_i) \end{bmatrix} \tag{11.1}$$

$J_i(\Phi_i) \in R^{6\times 6}$ between the two frames, the Jacobian transformation matrix is expressed as

$$
J_{1i}(\Phi_i) = \begin{bmatrix} \cos\psi_i \cos\theta_i & \sin\theta_i \sin\varphi_i \cos\psi_i - \cos\varphi_i \sin\psi_i \\ \sin\psi_i \cos\theta_i & \sin\theta_i \sin\varphi_i \sin\psi_i + \cos\varphi_i \cos\psi_i \\ -\sin\theta_i & \sin\varphi_i \cos\theta_i \end{bmatrix}
$$

$$
\begin{array}{c} \sin\varphi_i \sin\psi_i + \sin\theta_i \cos\varphi_i \cos\psi_i \\ \sin\theta_i \cos\varphi_i \sin\psi_i - \sin\varphi_i \cos\psi_i \\ \cos\varphi_i \cos\theta_i \end{array}
$$

$$
J_{2i}(\Phi_i) = \begin{bmatrix} 1 & \sin\varphi_i \tan\theta_i & \cos\varphi_i \tan\theta_i \\ 1 & \cos\varphi_i & -\sin\varphi_i \\ 0 & \sin\varphi_i \sec\theta_i & \cos\varphi_i \sec\theta_i \end{bmatrix}
$$

Remark 1: The predicate $|\theta_i(t)| < \pi/2$ guarantees that $J_i(\Phi_i)$ is a full-rank matrix because $\cos\theta_i$ is utilized as the determinant of $J_i(\Phi_i)$ in the equation above. This suggests that the angle θ_i should be set to obtain $-\dfrac{\pi}{2} < \theta < \dfrac{\pi}{2}$.

The initial AUV's dynamic model is used from Rani and Kumar (2023).

$$
M_i \dot{\chi}_i + C_i(\Phi_i)\chi_i + D_i(\chi_i)\chi_i + g_i(\chi_i) + T_{di} = \tau_i
$$

The ith AUV's centripetal matrix and coriolis, which include the extra mass components and rigid frame is represented by the matrix $C_i(\Phi_i) \in R^{6\times 6}$, on the other hand, the AUV's inertia matrix, which comprises the stiff body and additional mass terms, is represented by the matrix $M_i \in R^{6\times 6}$. $D_i(\Phi_i) \in R^{6\times 6}$, represents the hydrodynamic damping and lift matrix of the $i$thAUV. $g_i(\chi_i) \in R^{6\times 1}$ expressing the AUV's moments and gravitational forces. $T_{di} \in R^{6\times 1}$ is the source of the external disruption. $\tau_i = [X_i, Y_i, Z_i, L_i, M_i, N_i]^T$ is an application of generalized forces on the ith AUV supplied by the thrusters.

Following the kinematic conversion from (11.1) to (11.2) leads to the inertial frame AUV dynamics determination.

$$
\bar{M}_i \ddot{\Phi}_i + \bar{C}_i(\Phi_i, \chi_i)\dot{\Phi}_i + \bar{D}_i(\Phi_i, \chi_i)\dot{\Phi}_i + \bar{g}_i(\chi_i) + \bar{T}_{di} = \bar{\tau}_i \tag{11.2}
$$

where $\bar{M}_i = J_i^{-T}(\Phi_i)M_i J_i^{-1}(\Phi_i); \quad \bar{C}_i(\Phi_i, \chi_i) = J_i^{-T}(\Phi_i)[C_i(\chi_i) - M_i J_i^{-1}(\Phi_i)\dot{J}_i(\Phi_i)]$
$J_i^{-1}(\Phi_i); \bar{D}_i(\Phi_i, \chi_i) = J_i^{-T}(\Phi_i)D_i J_i^{-1}(\Phi_i); \bar{g}_i(\chi_i) = J_i^{-T}(\Phi_i)g_i; \bar{T}_d = J_i^{-T}(\Phi_i)T_{di}$

Property 1: The matrices $\bar{M}_i$ are both symmetric and positive definite.

Property 2: The Inertia and the Coriolis matrices satisfy the matrix skew-symmetric connection.

Property 3: Disturbances from outside are limited.

11.3 FORMULATION OF LEADER-FOLLOWER FORMATION

It is necessary for the leader AUV to follow the intended trajectory in the leader-following formation control problem, while the other follower AUVs must always maintain the desired angle and separation from the leader. Once every AUV has positioned itself correctly, the intended shaping effect is produced. When guiding a leader-follower formation, the follower AUV attempts to maintain the ideal separation while the leader AUV moves in the desired direction and comparative point from the leader. When every car is positioned as it should be, the development is at its best. Let's assume for the purposes of our discussion, that there is one leader (L) and two followers $(F_1$ and $F_2)$. The ideal distance d_{ij} can be used to achieve the general layout of the supporter. Let the intended position as well as the optimal direction's direction vector in case (A) and the optimal direction's direction vector in the instance (A) be represented by $\Phi_{di} = [x_i, y_i, z_i, \psi_i]^T$

The intended vector, tracking error, and filtered errors are defined as

$$\bar{\Phi}_i = \Phi_{di} - \Phi_i \tag{11.3}$$

Establish the filtered tracking error signal:

$$r_i = \dot{\bar{\Phi}}_{di} + K_{\Phi_i} \bar{\Phi}_i \tag{11.4}$$

where K_{Φ_i} is positive constant matrix. The desired velocity vector is then defined by:

$$J_i^{-1}(\Phi_{di})\dot{\Phi}_{di} = \chi_{di} \tag{11.5}$$

Using the aforementioned equation, the velocity tracking error is calculated.

$$\bar{\chi}_i = \chi_{di} - \chi_i \tag{11.6}$$

After applying (11.3) and differentiating (11.4), it is possible to establish the dynamic equation (11.2) in the form.

$$\bar{M}_i \dot{r}_i + \bar{C}_i r_i = \text{dynamics}_i - \bar{\tau}_i + \bar{T}_{di} \tag{11.7}$$

where $\text{dynamics}_i = \bar{M}_i \left(\dot{\bar{\Phi}}_{di} + K_{\Phi_i} \dot{\bar{\Phi}}_i \right) + \bar{C}_i \left(\dot{\Phi}_{di} + K_{\Phi_i} \bar{\Phi}_i \right) + \bar{D}_i \dot{\Phi}_i + \bar{g}_i$ and $u = [\bar{\Phi}_i, \dot{\bar{\Phi}}_i,$ $\Phi_{di}, \dot{\Phi}_{di}, \ddot{\Phi}_{di}]^T$ is the vector of input.

11.4 DESIGN OF CONTROLLER

The hydrodynamic loads that the AUVs are subjected to make it challenging to determine and measure the hydrodynamic coefficients necessary for maximum performance. In light of the work published in the literature, it's a prevalent misconception that the control system's dynamic knowledge is either entirely known or unknown.

It is hard to precisely determine every piece of dynamic information related to the dynamic system when there is any degree of nonlinearity or coupling. In this scenario, the controller's planning should make advantage of all accessible information on the framework's components. We shall segment the non-linear dynamic component in order to focus on addressing the issue into known $\left(\text{kndynamics}_i\right)$ and unknown $\left(\text{unkndynamics}_i\right)$ dynamic parts as

$$\overline{\text{kndynamics}_i} = \widehat{M}_i\left(\ddot{\Phi}_{di} + K_{\Phi_i}\dot{\Phi}_i\right) + \widehat{C}_i\left(\dot{\Phi}_{di} + K_{\Phi_i}\bar{\Phi}_i\right) + \widehat{D}_i\dot{\Phi}_i + \widehat{g}_i;$$

$$\overline{\text{unkndynamics}_i} = \widetilde{M}_i\left(\ddot{\Phi}_{di} + K_{\Phi_i}\dot{\Phi}_i\right) + \widetilde{C}_i\left(\dot{\Phi}_{di} + K_{\Phi_i}\bar{\Phi}_i\right) + \widetilde{D}_i\dot{\Phi}_i + \widetilde{g}_i$$

To do this, we apply the radial basis function neural network described by Park and Sandberg (1991) for the estimation of the unknown portion. Consequently, the approximate value of unknown dynamic part is as follows:

$$\overline{\text{unkndynamics}_i} = \bar{L}_i^T \vartheta_i\left(u_i\right) + \epsilon_i\left(u_i\right) \tag{11.8}$$

where $\bar{L} \in R^{r \times \bar{y}}$ is optimum weight matrix, u represents the input vector. $\epsilon\left(\cdot\right): R^{5\bar{q}} \to R^S$ illustrates the estimate inaccuracy and $\epsilon\left(u\right) < \epsilon_S$ for $\epsilon_S > 0$. $\vartheta\left(u\right)$ is the Gaussian feature mathematically set up as: $\vartheta_i\left(u_i\right) = \exp\left(\dfrac{-u_i - e_j^2}{2g_j^2}\right)$. Here e_j and g_j represent, respectively, the centers and widths of RBFNN. After substituting $\overline{\text{unkndynamics}_i}$ from (11.8) in (11.7), the new equation is as follows:

$$\bar{M}_i\dot{r}_i + \bar{C}_i r_i = \overline{\text{kndynamics}_i} + \bar{L}^T\vartheta_i\left(u_i\right) + \epsilon_i\left(u_i\right) + \bar{T}_{di} - \bar{\tau}_i \tag{11.9}$$

Next, for each AUV, we suggest the following hybrid controller as

$$\bar{\tau}_i = \overline{\text{kndynamics}_i} + A_{di}\rho_i + \widehat{\bar{L}}^T\vartheta_i\left(u_i\right) + \bar{\Delta} \tag{11.10}$$

where $\widehat{\bar{L}}^T$ is inserted as a result of the NN weights' approximation. $A_{di} = A_{di}^T$ and δ_i are, respectively, the gain matrix and the adaptive term. Here, to turn off the effects of the unsettling impacts and the NNs estimation mistake, δ_i is provided to the regulator part in the following manner.

Taking into account Property 3 and using $\epsilon_i\left(u_i\right) < \epsilon_S$, the following expression is produced:

$$\bar{T}_{di} + \epsilon_i\left(u_i\right) \leq y_1 + \epsilon_{Si} \tag{11.11}$$

The following steps are taken to obtain the adaptive compensator:

$$\bar{A}_i = y_1 + \epsilon_{Si} \tag{11.12}$$

which can be given in (11.13) in representative form as

$$\bar{A}_i = \begin{bmatrix} 1 & 1 \end{bmatrix} \begin{bmatrix} y_1 & \epsilon_{Si} \end{bmatrix}^T = K_i^T T_i \tag{11.13}$$

$$\bar{\Delta} = \frac{-\hat{\bar{B}}^2 r_i}{\hat{B} r_i + \bar{\Omega}_i}$$

In view of (11.11), (11.12), and (11.13) and using $\dot{\bar{\Delta}} = -\varphi\, \bar{\Delta}$ with $\bar{\Delta}(0) > 0$; a design constant created. Next, the error dynamics is transformed into the following form using the above equations in the form:

$$\bar{M}_i \dot{r}_i + \bar{C}_i r_i = -A_{di} r_i - \bar{\Phi}_i + \hat{\bar{L}}^T \vartheta_i(u_i) - \frac{\hat{\bar{B}}^2 r_i}{\hat{B} r_i + \bar{\Omega}_i} + \epsilon_i(u_i) + \bar{T}_{di} \tag{11.14}$$

For the weight network and the parameter vector, the accompanying adaptive control laws are:

$$\dot{\hat{\bar{L}}}_i = \Gamma_{L_i}\, \vartheta_i(u_i) r_i^T \tag{11.15}$$

$$\dot{\hat{T}}_i = \Gamma_{T_i}\, Z_i^T r_i \tag{11.16}$$

where $\Gamma_{L_i} = -\Gamma_{Li}^T \in R^{h_1 \times h_1}$ and $\Gamma_{\bar{X}} = \Gamma_{\bar{X}}^T \in R^{h_2 \times h_2}$; matrices with positive definiteness should be chosen.

Finally, the form in which the error dynamics is converted is:

$$\bar{M}_i \dot{r}_i + \bar{C}_i r_i = -A_{di} r_i - \bar{\Phi}_i + \hat{\bar{L}}^T \vartheta_i(u_i) - \frac{\hat{\bar{B}}^2 r_i}{\hat{B} r_i + \bar{\Omega}_i} + \epsilon_i(u_i) + \bar{T}_{di} \tag{11.17}$$

11.5   STABILITY ANALYSIS

To illustrate the stability of the overall structure and the resilience of the recommended control system, we characterize the Lyapunov characteristic as a feature of the following variables:

$$L_{lap_i} = \frac{1}{2} r_i^T \bar{M}_i r_i + \frac{1}{2} tr\left(\tilde{L}_i^T \Gamma_{L_i}^{-1} \tilde{L}_i\right) + \frac{1}{2} tr\left(\tilde{T}_i^T \Gamma_{\bar{X}}^{-1} \tilde{T}_i\right) + \frac{\overline{\Omega}_i}{\gamma_i} \tag{11.18}$$

Differentiation of equation (11.18) provides:

$$\dot{L}_{lap_i} = \dot{r}_i^T \bar{M}_i r_i + \frac{1}{2} r_i^T \dot{\bar{M}}_i r_i + tr\left(\tilde{L}_i^T \Gamma_{L_i}^{-1} \dot{\tilde{L}}_i\right) + tr\left(\tilde{T}_i^T \Gamma_{\bar{X}}^{-1} \dot{\tilde{T}}_i\right) + \frac{\dot{\overline{\Omega}}_i}{\gamma_i}$$

$$\dot{L}_{lap_i} \leq -2r_i^T B_d \dot{r}_i \tag{11.19}$$

Point out that $\dot{L}_{lap_i}$ converges to zero as $t \rightarrow \infty$ and the system as a whole is asymptotically stable by applying Barbalat's lemma.

Remark: The proposed control system guarantees that the formation errors fulfill preset performance standards and stabilizes them to small neighborhoods of origin, based on the results of the stability analysis presented above.

11.6   SIMULATION STUDY

This section illustrates the effectiveness of the proposed strategy to maintain control over the leader-follower formation in conjunction with AUVs utilizing MATLAB-based computational simulations. Here, the suggested method has been accomplished through the use of a 4-DOF simplified control mechanism.

This section uses numerical simulations in MATLAB to demonstrate the viability and effectiveness of the established plan for maintaining control over leader-follower formation when using AUVs. In this instance, we have implemented the suggested control strategy using a 4-DOF simplified AUV model, that is, controlling in x_i , y_i, z_i, and ψ_i directions as suggested by Rani and Kumar (2023). In this instance, we take into account p_i, q_i, ϕ_i, and θ_i all are zero. The updated version of the model is built in this manner. In this model, we have $\Phi_i = [x_i,\ y_i, z_i, \psi_i\ ]^T$ and $\chi_i = [u_i,\ v_i, w_i,\ t_i\ ]^T$. Also $M_i(\Phi_i) \in R^{4 \times 4}$; $C_i(\Phi_i) \in R^{4 \times 4}$ $D_i \in R^{4 \times 4}$ $g_i(\chi_i) \in R^{4 \times 1}$ and $T_{di} \in R^{4 \times 1}$ and these dynamic matrices are same as in Kumar and Rani (2020). To show the benefits, precision, and strength of the recommended control strategy, the outcomes of the simulation are applied to four formation control problems to investigate changes in the position and velocity errors.

Example (I) We take 2D control of the leader-follower formation into consideration. We take into account the following for the circular inertial planar preferred trajectory:

$$x_{Leader(d)} = 8\sin(0.01t);\ y_{Leader(d)} = 8\cos(0.01t)$$

The follower AUV's intended course in relation to the leader's configuration is therefore represented by the subsequent set of equations:

$$x_{follower1(d)} = x_{Leader(d)} + 10;\ y_{follower1(d)} = y_{Leader(d)} + 10$$

Using the model-based controller and the proposed hybrid control approach presented in (11.12), two distinct control strategies, we have produced simulation results for Case (I). these graphs, making it abundantly evident that, thanks to the efficient use of the available dynamic information, the proposed controller performs far better at tracking and formation control than the model-based controller (MBC).

Case (II) We further consider the control of 3D leader-follower formation. The following equations are then used to determine the optimal paths for the first and second AUVs to take after leaving the leader's setup:

$$x_{Leader(d)} = 40t + 10;\ y_{Leader(d)} = 30t - 5;\ Z_{Leader(d)} = 10$$

$$x_{follower1(d)} = 40t - 10; \quad y_{follower1(d)} = 30t + 5; \quad Z_{follower1(d)} = 20$$

$$x_{follower2(d)} = 40 + 30; \quad y_{follower2(d)} = 30t + 5; \quad Z_{follower2(d)} = 30$$

Comparative velocity errors in the x, y, and z directions are shown for the leader and its two followers in Figures 11.1–11.3. These figures demonstrate that the suggested control technique reduces the inaccuracies in velocity and positioning to zero levels in around 5 minutes when compared to the MBC. The controller that has been suggested has exhibited exceptional performance, proving its efficacy in controlling leader-follower formation. In conclusion, we discover that despite the neural network-based controller outperforming the adaptive-robust controller and MBC, the suggested hybrid controller really lessens the efficacy of the model-based control

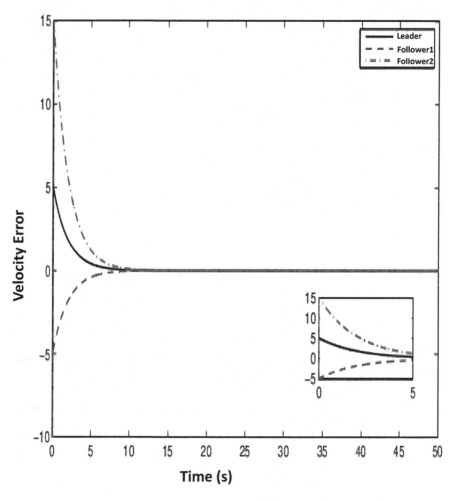

FIGURE 11.1 x-direction velocity errors using a recommended controller.

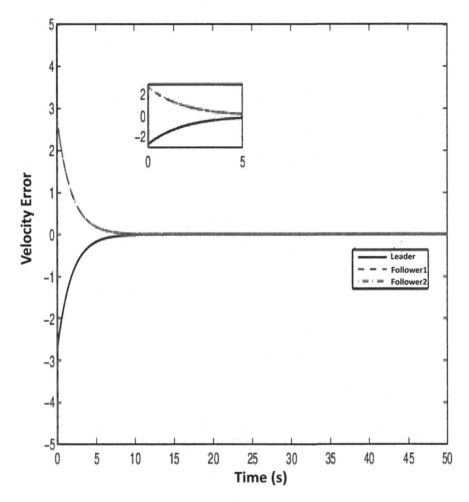

FIGURE 11.2 y-direction velocity errors using a recommended controller.

strategy by utilizing the neural network's approximation skills. This is based on the entire discussion. In the leader-follower formation control job, the recommended method performs better than the others and makes full use of the incomplete information supplied by the system dynamic model.

11.7 CONCLUSION

The formation control problem of many AUVs under unpredictable conditions is studied in this chapter. Initially, a thorough dynamic model that included numerous AUVs was created. More specifically, regarding the formation control issue, the controller design section included the model-free control schemes, the model-dependent control method, and the backstepping methodology. We have actively

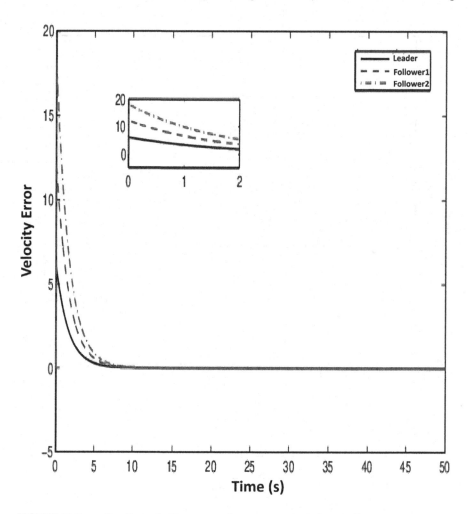

FIGURE 11.3 z-direction velocity errors using a recommended controller.

adjusted the unknown dynamic component using RBFNNs. To manage mistakes in network reconstruction and outside disturbances, the controller features an adaptive compensator. It takes into consideration the system's uncertainties as well. Afterward, the Lyapunov-based closed-loop stability synthesis covers the excellent and asymptotic converges of the formation position errors. Moreover, commanding a sizable formation of AUVs presents a number of challenges. They fall into three categories: barriers, collisions, and wave disturbances. The system's failure to take into consideration accidents, obstructions, or wave disturbances is one of the study work's weaknesses. To make the suggested system more resilient and assist in achieving the objective, all of these challenges might be incorporated into subsequent versions of the system.

REFERENCES

Atta, D., Subudhi, B., 2013. Decentralized formation control of multiple autonomous underwater vehicles. *Int. J. Robot. Autom.* 28 (4), 303–310.

Balch, T., Arkin, R.C., 1998. Behavior-based formation control for multirobot teams. *IEEE Trans. Syst. Man Cybern.: Syst.* 14 (6), 926–939.

Bechlioulis, C.P., Giagkas, F., Karras, G.C., Kyriakopoulos, K.J., 2019. Robust formation control for multiple underwater vehicles. *Front. Robot. AI* 6, 90.

Bejarbaneh, E.Y., Masoumnezhad, M., Armaghani, D.J., Pham, B.T., 2020. Design of robust control based on linear matrix inequality and a novel hybrid PSO search technique for autonomous underwater vehicle. *Appl. Ocean Res.* 101, 102231.

Breivik, M., Hovstein, V.E., Fossen, T.I., 2008. Ship formation control: A guided leader-follower approach. *IFAC Proc.* 41 (2), 16008–16014.

Elhaki, O., Shojaei, K., 2020. A robust neural network approximation-based prescribed performance output-feedback controller for autonomous underwater vehicles with actuators saturation. *Eng. Appl. Artif. Intell.* 88, 103382.

Emrani, S., Dirafzoon, A., Talebi, H., Nikravesh, S.Y., Menhaj, M., 2010. An adaptive leader-follower formation controller for multiple AUVs in spatial motions. In: *IECON 2010-36th Annual Conference on IEEE Industrial Electronics Society. IEEE*, Arizona, USA pp. 59–64.

Ge, S.S., Fua, C.H., Lim, K.W., 2004. Multi-robot formations: Queues and artificial potential trenches. In: *IEEE International Conference on Robotics and Automation, 2004. Proceedings. ICRA'04, 2004. IEEE*, New Orleans, LA, USA pp. 3345–3350.

Hou, S.P., Cheah, C., 2009. PD control scheme for formation control of multiple autonomous underwater vehicles. In: *2009 IEEE/ASME International Conference on Advanced Intelligent Mechatronics. IEEE*, Singapore pp. 356–361.

Huang, H., Tang, Q., Zhang, G., Zhang, T., Wan, L., Pang, Y., 2020. Multibody systembased adaptive formation scheme for multiple under-actuated AUVs. *Sensors* 20 (7), 1943.

Huang, Z., Zhu, D., Sun, B., 2016. A multi-AUV cooperative hunting method in 3-D underwater environment with obstacle. *Eng. Appl. Artif. Intell.* 50, 192–200.

Kumar, N., Rani, M., 2020. An efficient hybrid approach for trajectory tracking control of autonomous underwater vehicles. *Appl. Ocean Res.* 95, 102053.

Kumar, R., Stover, J.A., 2000. A behavior-based intelligent control architecture with application to coordination of multiple underwater vehicles. *IEEE Trans. Syst. Man Cybern.* 30 (6), 767–784.

Li, H., Yan, W., 2016. Model predictive stabilization of constrained underactuated autonomous underwater vehicles with guaranteed feasibility and stability. *IEEE ASME Trans. Mechatron.* 22 (3), 1185–1194.

Lu, Y., Wen, C., Shen, T., Zhang, W., 2020. Bearing-based adaptive neural formation scaling control for autonomous surface vehicles with uncertainties and input saturation. *IEEE Trans. Neural Netw. Learn. Syst.* 32 (10), 4653–4664.

Lu, Y., Xu, X., Qiao, L., Zhang, W., 2021. Robust adaptive formation tracking of autonomous surface vehicles with guaranteed performance and actuator faults. *Ocean Eng.* 237, 109592.

Mancilla, A., García-Valdez, M., Castillo, O., Merelo-Guervós, J.J., 2022. Optimal fuzzy controller design for autonomous robot path tracking using population-based metaheuristics. *Symmetry* 14 (2), 202.

vehicles subject to communication delays. *IEEE Trans. Control Syst. Technol.* 22 (2), 770–777.

Park, J., Sandberg, I.W., 1991. Universal approximation using radial-basis-function networks. *Neural Comput.* 3 (2), 246–257.

Peng, Z., Wang, J., 2017. Output-feedback path-following control of autonomous underwater vehicles based on an extended state observer and projection neural networks. *IEEE Trans. Syst. Man Cybern. Syst.* 48 (4), 535–544.

Peng, Z., Wang, D., Chen, Z., Hu, X., Lan, W., 2012. Adaptive dynamic surface control for formations of autonomous surface vehicles with uncertain dynamics. *IEEE Trans. Control Syst. Technol.* 21 (2), 513–520.

Peng, Z., Wang, D., Hu, X., 2011. Robust adaptive formation control of underactuated autonomous surface vehicles with uncertain dynamics. *IET Control Theory Appl.* 5 (12), 1378–1387.

Qu, Y., Xiao, B., Fu, Z., Yuan, D., 2018. Trajectory exponential tracking control of unmanned surface ships with external disturbance and system uncertainties. *ISA Trans.* 78, 47–55.

Rani, M., Kumar, N., 2023. A neural network based efficient leader-follower formation control approach for multiple autonomous underwater vehicles. *Eng. Appl. Artif. Intell.* 122, 106102.

Ren, W., Sorensen, N., 2008. Distributed coordination architecture for multi-robot formation control. *Robot. Auton. Syst.* 56 (4), 324–333.

Riahifard, A., Hosseini Rostami, S.M., Wang, J., Kim, H.-J., 2019. Adaptive leader-follower formation control of under-actuated surface vessels with model uncertainties and input constraints. *Appl. Sci.* 9 (18), 3901.

Sahoo, A., Dwivedy, S.K., Robi, P., 2019. Advancements in the field of autonomous underwater vehicle. *Ocean Eng.* 181, 145–160.

Shojaei, K., 2015. Leader-follower formation control of underactuated autonomous marine surface vehicles with limited torque. *Ocean Eng.* 105, 196–205.

Shojaei, K., 2016. Observer-based neural adaptive formation control of autonomous surface vessels with limited torque. *Robot. Auton. Syst.* 78, 83–96.

Tong, S., Wang, T., Tang, J.T., 2000. Fuzzy adaptive output tracking control of nonlinear systems. *Fuzzy Sets Syst.* 111 (2), 169–182.

Wang, J., Wang, C., Wei, Y., Zhang, C., 2019. Neuroadaptive sliding mode formation control of autonomous underwater vehicles with uncertain dynamics. *IEEE Syst. J.* 14 (3), 3325–3333.

Wang, J., Wang, C., Wei, Y., Zhang, C., 2020. Sliding mode based neural adaptive formation control of underactuated AUVs with leader-follower strategy. *Appl. Ocean Res.* 94, 101971.

Wang, Y., Yan, W., Li, J., 2012. Passivity-based formation control of autonomous underwater vehicles. *IET Control Theory Appl.* 6 (4), 518–525.

Xiang, X., Lapierre, L., Jouvencel, B., 2015. Smooth transition of AUV motion control: From fully-actuated to under-actuated configuration. *Robot. Auton. Syst.* 67, 14–22.

Xu, G., Jiang, W., Wang, Z., Wang, Y., 2022. Autonomous obstacle avoidance and target tracking of UAV based on deep reinforcement learning. *J. Intell. Robot. Syst.* 104 (4), 1–13.

Yan, Z., Wang, M., Xu, J., 2019. Integrated guidance and control strategy for homing of unmanned underwater vehicles. *J. Franklin Inst.* 356 (7), 3831–3848.

Yuan, J., Zhang, F.L., Zhou, Z.H., 2014. Finite-time formation control for autonomous underwater vehicles with limited speed and communication range. *Appl. Mech. Mater.* 511, 909–912.

Yun, B., Chen, B.M., Lum, K.Y., Lee, T.H., 2010. Design and implementation of a leaderfollower cooperative control system for unmanned helicopters. *J. Control Theory Appl.* 8 (1), 61–68.

12 Research Applications in Healthcare with Machine Learning Approaches

Vineeta Gulati and Arun Kumar Rana

12.1 INTRODUCTION

A branch of artificial intelligence called "machine learning" is concerned with creating statistical models and algorithms that let computer systems understand and a computer program is assigned specific tasks to accomplish, and if the machine's quantifiable performance on these tasks shows improvement as it gains more experience in carrying them out, it is said to have learned from its past experiences (Das et al., 2015). The machine utilizes data to make decisions and make predictions or forecasts. Consider the case of computer software that utilizes machine learning techniques to identify and forecast the presence of cancer based on the analysis of medical examination records pertaining to a patient. The performance of the system is expected to be enhanced with increased expertise, which will be achieved through the analysis of medical investigation reports from a broader population of patients. The performance of the system will be evaluated based on the accuracy of its predictions and its ability to detect cancer cases, which will be verified by an oncologist with extensive experience in the field (Ray & Chaudhuri, 2021). Robotics, virtual personal assistants (like Google), computer games, pattern recognition, natural language processing, data mining, traffic prediction, online transportation networks (like Uber, which estimates surge prices during peak hours), product recommendation systems, stock market prediction, medical diagnosis, online fraud detection, agricultural advisory services, search engine result refinement (like Google), chatbots for online customer support, email spam filtering, crime prediction: Three basic domains are usually covered by machine learning: classification, regression, and clustering. The availability of training data types and classifications influences the machine learning method selection. In such cases, one must consider approaches such as supervised learning, unsupervised learning, semi-supervised learning, and reinforcement learning to determine the most suitable approach (Figure 12.1).

12.1.1 Supervised Learning

Supervised learning is used to describe a machine learning technique where a model is trained with labeled data, each data point having a corresponding goal variable attached to it. The goal of the aforementioned learning method is characterized by

DOI: 10.1201/9781032673479-15

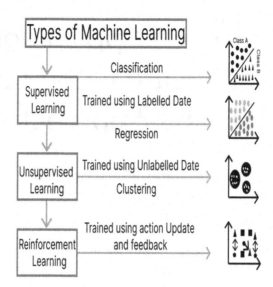

FIGURE 12.1 Types of machine learning.

its simplicity and ease of implementation (Sen et al., 2020). Supervised learning refers to a learning paradigm in which a dataset assumes the role of a teacher or guide throughout the training process of a model. In this methodology, the model undergoes automatic training and then engages in predictive and decision-making processes. Supervised learning primarily employs two distinct processes: The topic at hand pertains to the process of categorizing or grouping objects and ideas. In layman's terms, it refers to the anticipation of a categorical designation for a given data point. The process of classifying incoming data based on past data samples and categorizing them to train a model is referred to as classification.

The concept of regression refers to a statistical method used to model the relationship between a dependent variable. The task of predicting a numerical label, commonly referred to as regression, involves estimating a continuous output variable based on input data. Regression can be defined as the process of predicting continuous outcomes, identifying patterns, such as height and width, and understanding numbers and values.

12.1.2 Unsupervised Learning

Unsupervised learning describes a machine learning technique in which a model is trained on a dataset without the need for labeled instances or explicit supervision. In this paradigm, the model is tasked with unsupervised learning, which refers to the process through which a model endeavors to identify patterns within a given dataset, using its observations to discern underlying structures within the data (Muhamedyev et al., 2015). For instance, when photos of bananas, apples, and mangoes are provided to our model, the process of unsupervised learning involves identifying patterns and establishing relationships to generate clusters, dividing the dataset into distinct groups. When new data is sent to the model, it is automatically included in the existing cluster.

Clustering, in its simplest form, refers to the process of identifying groups inside a dataset. This technique involves exploring the internal structure of the data without any prior information, with the aim of organizing the data into meaningful clusters.

12.1.3 SEMI-SUPERVISED LEARNING

Semi-supervised learning refers to a machine learning approach that combines both labeled and unlabeled data for training models. The nomenclature of this method suggests that it falls within the spectrum of learning approaches, positioned between supervised and unsupervised learning. In certain scenarios, the cost of labeling is quite high, necessitating the presence of highly qualified individuals who can effectively deal with a limited number of labels due to their scarcity.

12.1.4 REINFORCEMENT LEARNING

This form of learning has distinct characteristics in comparison to both supervised and unsupervised learning approaches. In this educational context, a feedback loop is established between an autonomous agent and its surrounding environment. In order to establish a relationship or establish a connection between an agent and its environment, a set of actions is provided. The video game serves as a prime illustration of reinforcement. The process of reinforcement can be broken down into the following sequential steps:

- The agent was aware of the input state.
- Agents carry out actions based on instructions provided by their decision-making function.
- The agent executes an action and then receives reinforcement from the environment.
- The state-action information is retained for subsequent utilization.

12.1.5 CONTRIBUTION

- This chapter first focuses on the use of machine learning in the healthcare industry, and it examines previous research on the topic.
- Second, the healthcare decision support system is examined.
- Third, the role of machine learning's function in healthcare is explored.

This chapter is divided into the following sections: Section 12.2 describes the literature review. Sections 12.3 and 12.4 explore the decision support systems in healthcare and healthcare using machine learning, while Sections 12.5 and 12.6 define the role of machine learning in healthcare and the performance metrics used for classification systems. Conclusion and a prospective analysis for future developments are provided in Section 12.7.

12.2 LITERATURE REVIEW

Sr. No.	Reference	Concept Theoretical Model	Research Findings
1	Shailaja and Jabbar (2018)	The disease investigation is performed by presenting decision support, which helps in disease research by arranging an awareness of health issues or revealing the background details of individual patients in healthcare applications (heart disease, breast cancer, diabetes, and thyroid disease) in machine learning	Naive Bayes 86% (heart disease) SVM 96.40% (breast cancer) CART 79% (diabetic disease)
2	Ganiger and Rajashekharaiah (2018)	The focus question is to analyze chronic disease by machine learning algorithms with a data mining approach. Different datasets are available of distinct sizes and attributes for heart disease, liver disease, and diabetes to check the efficiency of machine learning algorithms. Decision tree (DT), support vector machine (SVM), and Random forest (RF) algorithms are used, and Random Forest proved to be the best for heart and diabetes disease	Heart disease 83% Diabetic 98%
3	Ray and Chaudhuri (2021)	Presented machine learning algorithms from a healthcare perspective implementation, their merits, and demerits in terms of performance	Comparisons of the machine learning algorithm are shown based on different performance metrics (accuracy, confusion matrices, etc.) related to machine learning technology along with their advantages and disadvantages
4	Tekale et al. (2018)	Here, the work is to make a model with the highest precision for predicting whether CKD is correct or not. Based on age, gender, blood, and creatinine level, calculate GFR to find out the level of kidney function	SVM algorithm implementation provides the best accuracy of 96.50%

(Continued)

(Continued)

Sr. No.	Reference	Concept Theoretical Model	Research Findings
5		Presented the role of machine learning algorithms in various healthcare sectors like diagnosis of disease, brain computing, biometrics, etc. Also discussed deep learning and its role in disease identification	Through machine learning and deep learning performed well in all categories of the healthcare sector, but some constraints were faced when we implemented this technology like regulations lack due to the safety or efficiency of algorithms. Another issue that is usually faced is the sharing of data in the current healthcare sector
6	Fatima and Pasha (2017)	Here, the focus question is to compare all the machine learning algorithms for the identification of distinct chronic diseases related to heart, kidney, cancer, etc., analyze them, and make the decision accordingly	After the implementation process of all the algorithms, Diabetes disease is accurately identified by the naive Bayes algorithm with an accuracy of 95%. In the same way, the neural network algorithm gives the best result in the case of hepatitis disease, etc.

12.3   DECISION SUPPORT SYSTEMS IN HEALTHCARE

A significant number of individuals in the United States succumb annually due to deficiencies within the healthcare system, while a considerable population has a decline in health as a result of nonfatal burns caused by persistent factors. The Health Information Technology (IT) Framework proposes several techniques, including collaboration, understanding consumer preferences for physicians and organizations, and the implementation of IT (Sebaa et al., 2017).

The concept of decision support refers to the provision of tools and resources that assist individuals or organizations in making informed and effective decisions. A health protection system that utilizes machine learning effectively leverages the computational power of computers and the cognitive abilities of doctors. Both machines and doctors engage in the process of pattern recognition (Hamedan et al., 2020). However, doctors are limited in their ability to assess the heartbeat of every patient or possess comprehensive knowledge of the intricacies associated with each condition. The machine will do the aforementioned activities and thereafter give its results to the physician for verification.

The application of decision support systems in the healthcare industry The Decision Support System assists the financial department of the clinic in monitoring charges, accounts receivable, expenditures, and accounts payable. Furthermore, this approach aids in maintaining the patient's insurance coverage and provides many options for reimbursement. The firm offers a diverse range of modules for decision support systems (DSS) in the healthcare sector (Hasan et al., 2017). The utilization of DSS in disease investigation facilitates the provision of a comprehensive understanding of health issues to medical professionals, as well as the dissemination of pertinent background information pertaining to particular patients. Furthermore, it offers assistance in the identification of a patient's condition and provides guidance to the patient regarding the appropriate timing for administering the correct medication. This is facilitated through the utilization of a web-based framework integrated with an automated patient/therapeutic record system.

12.4 HEALTHCARE USING MACHINE LEARNING

The advancement of health-related information has presented significant opportunities for enhancing patient recovery. Machine learning has emerged as a crucial tool in healthcare, particularly in areas such as computer-aided diagnosis, image registration, image annotation, image-guided medical interventions, and image database retrieval (Araújo et al., 2016). Additionally, machine learning techniques have been applied to multimodal image fusion and medical image segmentation, addressing challenges where conventional approaches may be ineffective in achieving a cure. The potential social consequences of machine learning in the healthcare industry may be rather limited. Machine learning offers a potential answer for mitigating the rising costs of healthcare and facilitating enhanced communication between patients and clinicians. Machine learning (ML) solutions are increasingly being employed in many health-related applications. One such application involves assisting physicians in identifying more personalized prescriptions and therapies for their patients (Kumar et al., 2021). Additionally, ML can aid patients in determining the appropriate timing and necessity of scheduling follow-up appointments. In the field of healthcare, there has been a significant increase in the availability of a vast amount of information. The system encompasses electronic medical records that comprise both organized and unstructured information. Structured health information refers to data that is easily analyzable within a database. This data includes a variety of metrics and classifications, such as patient weights and common symptoms like headaches and upset stomachs. Unstructured data in a variety of formats, such as notes, images, reports, audio and video recordings, and discharge summaries, make up the majority of medical knowledge. Quantifying and analyzing a dialogue between a provider and a patient can be a challenging task due to its highly individualized nature and the potential for diverse trajectories.

12.4.1 ETHICS OF USING ALGORITHMS IN HEALTHCARE

It has been previously asserted that the cognitive capacity of medical practitioners serves as the most effective ML tool in the field of healthcare. There was a period during which automobile industry workers harbored concerns regarding the potential displacement of their employment due to the advent of robotics (Tikariha & Richhariya, 2018). Likewise, certain medical practitioners may harbor concerns over the potential obsolescence that could arise from the advent of ML. However, the practice of medicine possesses an irreplaceable artistic quality. The presence of human interaction and the establishment of empathetic and compassionate connections between patients and healthcare providers will remain an essential aspect of patient care. ML, along with other forthcoming technologies in the field of medicine, will not eradicate this issue entirely. Instead, they will serve as instruments that healthcare professionals employ to enhance the provision of continuous medical treatment.

The primary emphasis should be placed on the utilization of ML techniques to enhance the provision of healthcare to patients. For instance, when doing cancer diagnostics on a patient, it is imperative to obtain biopsy results of the utmost quality. The development of a ML system capable of analyzing pathology slides and providing diagnostic assistance to pathologists holds significant value. If the results can be obtained in a significantly shorter duration while maintaining the same level of precision, it is expected that this will ultimately enhance both patient care and satisfaction. I include this personal anecdote as my own mother has been eagerly anticipating her test results for more than a week. The healthcare industry should transition its perspective on ML from a speculative notion to a practical instrument that can be readily implemented in the present. In order for ML to be integrated into the healthcare sector, it is imperative to adopt a gradual approach. It is imperative to identify certain instances when the capabilities of ML offer valuable contributions inside a distinct technical application, such as the example of Google and Stanford. This document outlines a systematic approach for integrating analytics, ML, and prediction algorithms into routine clinical practice.

In the first instance, it is imperative that our objectives align with our capacities. The process of training an ML algorithm to discern skin cancer from an extensive collection of skin cancer photographs is widely comprehended by the majority of individuals. If it were to be discovered that radiologists are being substituted by algorithms, individuals would likely exhibit a reasonable degree of reluctance (Jiang et al., 2017). The bridging process must be implemented gradually over a period of time. The obsolescence of radiologists is quite unlikely; nonetheless, in the future, radiologists will assume a supervisory role in reviewing and overseeing readings that have been first interpreted by automated systems. ML will be utilized as a collaborative partner to identify and highlight specific areas of attention, reduce noise, and assist in prioritizing areas of concern with a higher chance of significance. The field of medicine employs a systematic approach to examine and establish the safety and efficacy of various treatments. The process of trial and error is often protracted, including the systematic evaluation of various options and making conclusions based on empirical facts (Callahan & Shah, 2017). It is imperative to establish equivalent protocols when considering the implementation of ML in order to guarantee its safety and effectiveness. It is vital to comprehend the ethical considerations associated with delegating a portion of our tasks to an automated system.

12.5 ROLE OF MACHINE LEARNING IN HEALTHCARE

The application of ML in the early stages of drug development has significant promise for a diverse array of functions, encompassing initial drug screening as well as the prediction of success rates contingent upon biological variables (Ahamed & Farid, 2018). Furthermore, it can contribute to the anticipation of the efficacy and safety of the pharmaceutical compound, which constitutes a primary objective in the process of novel drug development. Additionally, it can furnish insights into potential novel constituents that hold promise as pharmaceutical agents. The necessity for this novel technique in drug development arises from the existing assumption under current standards that each chemical possesses a distinct purpose and undergoes rigorous professional investigation, hence constraining the scope of medication development to a limited selection of compounds. Furthermore, it is assumed that the initial combination to advance to clinical trials is considered the most optimal (Figure 12.2).

The diagnostic phase holds significant importance in patient follow-up during the course of care. ML offers healthcare providers novel time-saving options and improves the precision of diagnoses. Furthermore, it facilitates the exploration of novel concepts in the realm of disease diagnosis and prediction, encompassing a wide range of ailments. For example, the diagnosis of cardiac disease is currently being facilitated by the development of ML techniques by scientists. The automated system for diagnosing heart illness has been the focus of extensive global study and is anticipated to hold significant historical significance in the 21st century (Winter, 2019).

The prognosis of diabetes includes the potential for detrimental effects on multiple bodily systems, including but not limited to the cardiovascular system, renal system, and brain system. The early detection of diabetes using mechanical investigations is crucial for the preservation of patients' life. Various algorithms

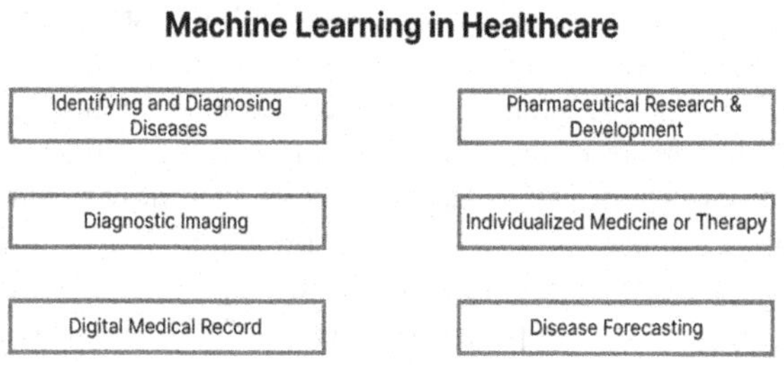

Machine Learning in Healthcare

Identifying and Diagnosing Diseases	Pharmaceutical Research & Development
Diagnostic Imaging	Individualized Medicine or Therapy
Digital Medical Record	Disease Forecasting

FIGURE 12.2 Application of ML in healthcare.

can be employed by healthcare practitioners to forecast the occurrence of diabetes (Jayatilake & Ganegoda, 2021). The utilization of data mining and ML techniques has recently been introduced in the development of a liver disease prediction system. The prediction of liver illness poses a significant problem due to the multitude of potential diseases that might impact the liver, as well as the extensive body of information available on the subject matter. Nevertheless, researchers are diligently endeavoring to tackle these concerns. Cancer screening encompasses a range of mechanical investigations that are capable of detecting and diagnosing various types of cancer. Since the advent of artificial intelligence (AI), researchers have made significant advancements in the development of intricate algorithms capable of identifying visual tissue and providing diagnoses for various forms of cancer. The recent advancements in AI have facilitated the precise identification of tissue, free from potential confounding factors such as air bubbles or bleeding, hence enhancing the accuracy of medical diagnoses (Bhardwaj et al., 2017). CAT scans, MRIs, and other imaging technologies yield voluminous data that surpasses the resolution of megapixels, posing a considerable challenge even for seasoned radiologists and pathologists. The field of statistics in the realm of healthcare is primarily concerned with enhancing the overall quality of the care provided to individuals. The utilization of data measurement and patient analytics plays a crucial role in facilitating informed decision-making within the medical field, as it furnishes empirical proof to substantiate medical data. There exist several distinct categories of analytics skills. Descriptive analyses and diagnoses are conventional methodologies that utilize historical data to provide statistical reports on the proportion of individuals afflicted with influenza, in comparison to the preceding and antecedent months.

12.6 PERFORMANCE METRICS USED IN ML FOR HEALTHCARE CLASSIFICATION SYSTEM

Following the deployment of the ML model, it is necessary to conduct testing in order to assess the efficacy of the model. This evaluation is typically performed using matrices, which may vary depending on the specific task being addressed by the ML algorithms. When assessing the performance of an ML model, it is imperative to choose the appropriate database for evaluating its performance. In the context of ML training, if the model's performance is solely assessed using the training data, it may exhibit bias in its output. This bias arises from the model's reliance on the training set during the training process. In order to assess the generalization error, it is necessary to analyze the ML model with data that has not been previously seen by the model. Hence, it is advisable to assess the performance of the ML model on the test dataset.

Metric	Example
Confusion Matrix: It is a matrix that consists of the table that is mainly used to describe the performance of the classification model	Actual Positive and Negative Data

	Positive	Negative
	60	30
	40	50

Predicted Positive and Negative Data
So out of a total of 100 positive samples, 60 are correctly classified as positive and the remaining 40 are misclassified as negative

Accuracy Measure:

True Positive: These are the cases that are correctly classified as yes or have the disease

True Negative: These are the cases that are correctly classified as no or don't have the disease

False-Positive: Here, the model predicted yes, but the patient doesn't have the disease

False Negative: Here, the model predicted no, but the patient has diseases

Accuracy: $(TP + TN)/(TP + TN + FP + FN)$

	Actual Cancer = Yes	Actual Cancer = No
Predicted Cancer = Yes	TP	FP
Predicted Cancer = No	FN	TN

12.7   CONCLUSION

ML approaches play a vital role in various domains of the business sector. The healthcare industry is currently encountering a multitude of challenges, resulting in an escalation in costs. This chapter presents the various ML approaches to address these issues. In contemporary times, the prevalence of kidney failure, lung diseases, and heart disease has significantly increased, emerging as prominent global health concerns. It is imperative to recognize chronic diseases at an early stage to prevent functional deterioration. In order to facilitate the prediction of chronic diseases among patients, a Chronic Disease Prediction System will be developed, aiming to assist doctors and medical specialists in their diagnostic endeavors. The primary emphasis lies in the utilization of ML algorithms for the early detection of chronic diseases, aiming to achieve enhanced accuracy.

REFERENCES

Ahamed, F., & Farid, F. (2018). Applying Internet of Things and machine-learning for personalized healthcare: Issues and challenges. In *2018 International Conference on Machine Learning and Data Engineering (ICMLDE)*, Sydney, Australia, pp. 19–21. https://doi.org/10.1109/iCMLDE.2018.00014.

Araújo, F. H. D., Santana, A. M., & de A. Santos Neto, P. (2016). Using machine learning to support healthcare professionals in making preauthorisation decisions. *International Journal of Medical Informatics*, *94*, 1–7. https://doi.org/10.1016/j.ijmedinf.2016.06.007.

Bhardwaj, R., Nambiar, A. R., & Dutta, D. (2017). A study of machine learning in health-care. *Proceedings - International Computer Software and Applications Conference*, *2*, 236–241. https://doi.org/10.1109/COMPSAC.2017.164.

Callahan, A., & Shah, N. H. (2017). Machine learning in healthcare. In Aziz Sheikh, David W. Bates, Adam Wright, Kathrin Cresswell (Eds) *Key Advances in Clinical Informatics: Transforming Health Care through Health Information Technology*. Elsevier Inc., Massachusetts, US, pp. 279–291. https://doi.org/10.1016/B978-0-12-809523-2.00019-4.

Das, S., Dey, A., & Roy, N. (2015). Applications of artificial intelligence in machine learning: Review and prospect. *International Journal of Computer Applications*, *115*(9), 31–41.

Fatima, M., & Pasha, M. (2017). Survey of machine learning algorithms for disease diagnostic. *Journal of Intelligent Learning Systems and Applications*, *09*(01), 1–16. https://doi.org/10.4236/jilsa.2017.91001.

Ganiger, S., & Rajashekharaiah, K. M. M. (2018). Chronic diseases diagnosis using machine learning. In *2018 International Conference on Circuits and Systems in Digital Enterprise Technology, ICCSDET 2018*, Kerala, India. pp. 1–6. https://doi.org/10.1109/ICCSDET.2018.8821235.

Hamedan, F., Orooji, A., Sanadgol, H., & Sheikhtaheri, A. (2020). Clinical decision support system to predict chronic kidney disease: A fuzzy expert system approach. *International Journal of Medical Informatics*, *138*, 104134. https://doi.org/10.1016/j.ijmedinf.2020.104134.

Hasan, M. S., Ebrahim, Z., Wan Mahmood, W. H., & Ab Rahman, M. N. (2017). Decision support system classification and its application in manufacturing sector: A review. *Jurnal Teknologi*, *79*(1), 153–163. https://doi.org/10.11113/jt.v79.7689.

Jayatilake, S. M. D. A. C., & Ganegoda, G. U. (2021). Involvement of machine learning tools in healthcare decision making. *Journal of Healthcare Engineering*, pp. 1–20 *2021*. https://doi.org/10.1155/2021/6679512.

Jiang, F., Jiang, Y., Zhi, H., Dong, Y., Li, H., Ma, S., Wang, Y., Dong, Q., Shen, H., & Wang, Y. (2017). Artificial intelligence in healthcare: Past, present and future. *Stroke and Vascular Neurology*, *2*(4), 230–243. https://doi.org/10.1136/svn-2017-000101

Kumar, N., Narayan Das, N., Gupta, D., Gupta, K., & Bindra, J. (2021). Efficient automated disease diagnosis using machine learning models. *Journal of Healthcare Engineering*, pp.1–13, *2021*. https://doi.org/10.1155/2021/9983652.

Muhamedyev, R. I., Pushkina, K., & Kazakhstan, A. (2015). Machine learning methods: An overview *Computer Modelling & New Technologies* 19(6), 14–29. www.cmnt.lv.

Ray, A., & Chaudhuri, A. K. (2021). Smart healthcare disease diagnosis and patient management: Innovation, improvement and skill development. *Machine Learning with Applications*, *3*, 100011. https://doi.org/10.1016/j.mlwa.2020.100011.

Sebaa, A., Nouicer, A., Tari, A., Tarik, R., & Abdellah, O. (2017). Decision support system for health care resources allocation. *Electronic Physician*, *9*(6), 4661–4668. https://doi.org/10.19082/4661.

Sen, P. C., Hajra, M., & Ghosh, M. (2020). Supervised classification algorithms in machine learning: A survey and review. *Advances in Intelligent Systems and Computing*, pp. 99–111 937. https://doi.org/10.1007/978-981-13-7403-6_11.

Shailaja, K., Seetharamulu, B. and Jabbar, M.A., 2018, March. Machine learning in healthcare: A review. In 2018 Second international conference on electronics, communication and aerospace technology (ICECA) (pp. 910-914). IEEE. Coimbatore, India.

Tekale, S., Shingavi, P., Wandhekar, S., & Chatorikar, A. (2018). Prediction of chronic kidney disease using machine learning algorithme. *International Journal of Advanced Research in Computer and Communication Engineering*, *7*(10), 92–96. https://doi.org/10.17148/IJARCCE.2018.71021.

Tikariha, P., & Richhariya, P. (2018). Comparative study of chronic kidney disease prediction using different classification techniques. *Lecture Notes in Networks and Systems*, 34, 195–203. https://doi.org/10.1007/978-981-10-8198-9_20.

Winter, G. (2019). Machine learning in healthcare. *British Journal of Health Care Management*, *25*(2), 100–101. https://doi.org/10.12968/bjhc.2019.25.2.100.

Section 4

Case Studies/Uses of IoT and Big Data in Different Domains

13 Enhancing Security, Privacy, and Predictive Maintenance through IoT and Big Data Integration

Debosree Ghosh

13.1 INTRODUCTION

The Internet of Things (IoT) and Big Data have emerged in recent years as revolutionary forces in a variety of industries as a result of the quick development of technology. Huge amounts of data have been produced by the proliferation of connected devices in the IoT, and Big Data technologies provide strong tools for analyzing and deriving insights from these data streams. Big Data and IoT's merger offers a singular opportunity to address pressing problems [1] with security, privacy, and preventive maintenance.

13.1.1 BACKGROUND

IoT refers to the network of physical devices, vehicles, appliances, and other items embedded with sensors, software, and connectivity to exchange data with other devices and systems over the internet. This connectivity enables real-time data collection, monitoring, and control of various processes. Meanwhile, Big Data encompasses the technologies and techniques for processing and analyzing large and complex datasets that are beyond the capabilities of traditional data management tools.

13.1.2 MOTIVATION

The way that organizations operate, innovate, and tackle problems might be completely transformed with the help of the integration of IoT-generated data and Big Data analytics. Through proactive threat detection, robust data anonymization techniques, and streamlined predictive maintenance procedures using real-time data insights, this integration can result in improved security measures.

13.1.3 SCOPE AND OBJECTIVES

This chapter aims to provide a comprehensive overview of the ways in which the integration of IoT and Big Data can enhance security, privacy, and predictive maintenance strategies. By examining the methodologies, technologies, and real-world

DOI: 10.1201/9781032673479-17

applications, we intend to demonstrate how this integration is reshaping industries and generating new opportunities for innovation.

13.2 LITERATURE REVIEW

The integration of IoT and Big Data has emerged as a transformative force across various industries, offering unprecedented opportunities to enhance security, privacy, and predictive maintenance. In this literature review, we delve into the existing body of knowledge to explore the intersection of IoT and Big Data, specifically focusing on their collective impact on security, privacy, and predictive maintenance.

The proliferation of interconnected devices in IoT ecosystems has brought about new challenges and opportunities for security enhancement. Various studies emphasize the importance of securing IoT devices to prevent unauthorized access, data breaches, and cyber-attacks. Researchers have explored encryption techniques, secure communication protocols, and anomaly detection algorithms to fortify the security of IoT networks. Notable works by authors such as [2] and [3] highlight the ongoing efforts to establish robust security frameworks within the realm of IoT.

As the deployment of IoT devices becomes ubiquitous, the preservation of user privacy becomes a paramount concern. Literature in this domain has focused on the development of privacy-preserving mechanisms, including anonymization techniques, user-centric control models, and edge computing solutions. Works by [4] and [5] shed light on the evolving landscape of privacy considerations in IoT, stressing the need for a balance between data utility and individual privacy.

The integration of Big Data analytics in predictive maintenance has revolutionized traditional approaches, enabling proactive and data-driven strategies. Extensive research has been conducted on leveraging machine learning algorithms, sensor data, and historical records to predict equipment failures and optimize maintenance schedules. Key contributions by [6] and [7] underscore the significance of data-driven predictive maintenance in enhancing equipment reliability and reducing downtime.

The convergence of IoT and Big Data presents a synergistic approach to addressing security, privacy, and predictive maintenance challenges. Studies by [8] and [9] explore the seamless integration of IoT-generated data into Big Data analytics platforms, emphasizing the potential for real-time threat detection, personalized privacy controls, and enhanced predictive maintenance models.

13.3 IOT AND BIG DATA INTEGRATION

The integration of IoT and Big Data technologies represents a pivotal transformation in how data is collected, processed, and leveraged for insights. This section delves into the mechanics and significance of merging IoT-generated data with advanced Big Data analytics.

IoT devices generate vast amounts of data through sensors and cameras, providing real-time insights into processes, surroundings, and user behaviors. Big Data technologies manage and analyze these complex datasets, enabling businesses to harness real-time data streams and use advanced analytics methods. IoT devices collect data from various sources, and Big Data technologies enable efficient data capture and storage. Big Data analytics tools enable real-time analysis of incoming data streams, enabling informed decisions on the fly. As IoT devices grow, the volume of generated data increases exponentially, and Big Data solutions provide scalable storage architectures. Predictive modeling and machine learning techniques can be applied to IoT data [10], enhancing predictive maintenance strategies. The integration of IoT and Big Data unlocks potential for innovation, improved decision-making, and a competitive edge.

13.4 ENHANCING SECURITY THROUGH INTEGRATION

The integration of IoT and Big Data technologies offers organizations a new security frontier. By integrating IoT-generated data with Big Data analytics, organizations can enhance their security measures, providing real-time insights and prompt responses to potential risks. Anomaly detection and predictive analytics, based on machine learning algorithms trained on IoT data, enhance proactive security solutions. This synergy equips businesses with the tools to protect sensitive data and assets in a digital world that is becoming increasingly interconnected.

The integration of IoT-generated data with Big Data analytics offers organizations real-time insights into their operational landscapes, enabling timely responses to security incidents and enhancing situational awareness. This holistic view of ongoing activities aids in identifying potential threats. Anomaly detection is crucial for robust security systems, and the combination of IoT and Big Data enhances this capability [11]. Historical data analysis using Big Data analytics establishes baseline patterns of normal behavior, flagging deviations as anomalies. Predictive security measures are adopted, with machine learning algorithms identifying subtle precursors to security incidents. The combination of IoT data and Big Data analytics creates comprehensive profiles of user and device behaviors, triggering alerts for potential security breaches or unauthorized access and enhancing threat detection precision.

13.5 SAFEGUARDING PRIVACY IN IOT AND BIG DATA CONTEXTS

Protecting individual privacy is a vital requirement in the constantly changing world of IoT and Big Data integration. Large-scale IoT data generation combined with potent Big Data analytics skills results in both complex privacy issues and previously unimaginable discoveries. A diverse strategy is necessary to safely manage this environment. By adding controlled noise to datasets, techniques like differential privacy enable accurate analysis while hiding individual contributions. By obscuring personally identifying information, data anonymization protects privacy in the event of breaches. Transparency and permission from consumers lay an ethical foundation, while strong encryption enables secure data transit and storage [12] and [13]. Accepting the idea of privacy by design allows solutions for IoT and Big Data to be built with security features already in place.

Strong privacy safeguards are required because of the massive volumes of personal data being collected as a result of the widespread use of IoT devices and Big Data analytics. The essential strategies for striking a balance between data value and privacy are encryption technologies, data anonymization, and differential privacy. Differential privacy enables relevant analysis while masking individual contributions from datasets by adding calibrated noise. Data anonymization obscures or encrypts information to safeguard individual identity. It is imperative to guarantee data security with strong encryption techniques and access controls. Transparency and user permission are the cornerstones of privacy preservation. Privacy by design incorporates privacy concerns into Big Data and IoT platforms (Figure 13.1).

13.6   IOT-DRIVEN PREDICTIVE MAINTENANCE

The integration of IoT and Big Data technologies has revolutionized the domain of predictive maintenance. This section delves into the transformative potential of IoT-driven predictive maintenance, exploring how real-time data, advanced analytics, and machine learning are reshaping maintenance practices across industries.

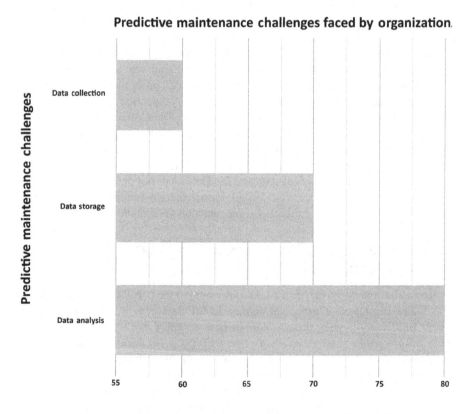

FIGURE 13.1   Predictive maintenance challenges faced by Organizations

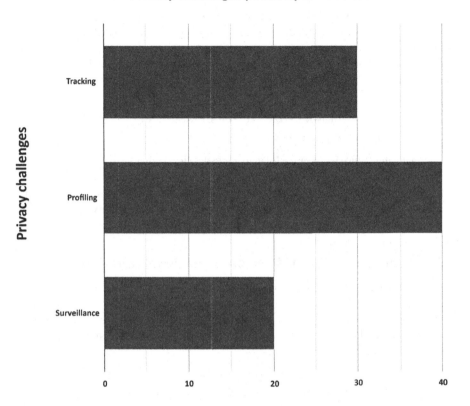

FIGURE 13.2   Privacy challenges posed by IoT devices

IoT-driven predictive maintenance uses real-time data streams from sensors embedded in machinery and equipment to anticipate potential failures and optimize maintenance schedules. This shift allows organizations to make informed decisions about maintenance tasks, minimizing downtime and maximizing operational efficiency. Machine learning algorithms, trained on historical IoT data, identify patterns and deviations, enhancing predictions over time. This proactive approach reduces operational costs and increases equipment uptime by resolving problems early and optimizing maintenance plans based on actual equipment conditions (Figure 13.2).

13.7   CASE STUDIES AND REAL-WORLD APPLICATIONS

The IoT and Big Data technological confluence has spurred a wave of ground-breaking applications across several industries. This section covers detailed case studies and real-world examples that demonstrate the practical advantages of this combination.

13.7.1 SMART CITIES: URBAN EFFICIENCY AND SAFETY

IoT sensors collect information on traffic patterns, trash disposal, and public utilities in the context of smart cities. This data is processed by Big Data analytics to enhance energy efficiency, enhance garbage collection routes, and optimize traffic patterns. Predictive policing is made possible through real-time analysis, which helps law enforcement organizations locate prospective crime hotspots and use their resources wisely.

13.7.2 MANUFACTURING: PREDICTIVE MAINTENANCE AND QUALITY ENHANCEMENT

Manufacturing industries leverage IoT-generated data for predictive maintenance. Sensors embedded in machinery continuously monitor parameters like temperature, vibration, and performance metrics. Big Data analytics analyze this data to predict maintenance needs accurately, minimizing downtime. Additionally, real-time quality control is fortified by monitoring production processes and ensuring adherence to quality standards.

13.7.3 HEALTHCARE: REMOTE MONITORING AND PERSONALIZED TREATMENT

IoT devices, including wearable health trackers and medical sensors, enable remote patient monitoring. Vital signs, medication adherence, and health trends are continuously collected. Big Data analytics analyze this data to provide personalized treatment recommendations, aiding healthcare providers in delivering proactive care and reducing hospitalizations.

13.7.4 ENERGY MANAGEMENT: EFFICIENCY AND SUSTAINABILITY

IoT devices track energy consumption patterns in buildings and industrial facilities. Big Data analytics analyze this data to identify usage trends and energy inefficiencies. Organizations can then optimize consumption, reduce costs, and contribute to sustainable practices by minimizing their carbon footprint.

13.7.5 AGRICULTURE: PRECISION FARMING AND CROP OPTIMIZATION

IoT sensors deployed in agricultural fields capture data on soil moisture, weather conditions, and crop health. Big Data analytics process this information to offer insights into optimal irrigation schedules, pest management strategies, and crop growth patterns. This integration facilitates precision farming, boosting yields and minimizing resource wastage.

13.7.6 TRANSPORTATION: LOGISTICS AND FLEET MANAGEMENT

IoT devices in vehicles provide real-time data on location, fuel consumption, and vehicle health. Big Data analytics analyze this data to optimize routes, enhance fuel efficiency, and streamline fleet management operations. This integration improves logistics, reduces costs, and contributes to greener transportation practices.

13.7.7 RETAIL: CUSTOMER INSIGHTS AND PERSONALIZATION

Retailers utilize IoT-generated data from in-store beacons, online interactions, and purchase history to understand customer behavior. Big Data analytics analyze this data to create personalized shopping experiences, targeted promotions, and optimized inventory management. This integration enhances customer engagement and drives revenue growth.

13.8 OVERCOMING CHALLENGES AND ADDRESSING RISKS

Big Data and IoT's synergy has great promise, but there are many obstacles to overcome and pitfalls to be aware of. To protect user privacy, organizations must prioritize data security through encryption and user permission. Data volume issues are addressed by effective data management techniques, such as scalable storage systems. Complexities in integration necessitate using agile methods and established protocols. Validation and normalization play a role in ensuring data quality. Training can close skill gaps and produce competent workers. Responsible practices are maintained through ethical concerns and regulatory compliance [14]. System uptime is avoided by achieving redundancy and maintenance. Shared knowledge and collaboration promote communal development. Organizations may fully leverage the transformative potential of IoT and Big Data integration by overcoming these obstacles.

The rise of IoT devices has led to a significant increase in sensitive data, necessitating robust security measures. Organizations must implement encryption, secure authentication, and regular audits to protect data from breaches. Privacy concerns should be addressed through data anonymization, user consent, and data protection regulations. The data influx presents challenges in storage, processing, and management. Scalable cloud infrastructures, distributed storage solutions, and data pruning techniques can help manage large datasets. Integrating diverse IoT devices and Big Data platforms can be complex, but standardized protocols and agile approaches can facilitate communication. Data validation, cleaning, and normalization processes ensure accurate and reliable data for Big Data analytics. Skilled workforce in data analysis, machine learning, and cybersecurity is essential. Reliable operation of IoT and Big Data systems is crucial. Collaboration among stakeholders is essential for overcoming challenges and mitigating risks.

13.9 ETHICAL CONSIDERATIONS IN IOT AND BIG DATA INTEGRATION

A new era of data-driven innovation has begun as a result of the convergence of IoT and Big Data technology. However, this integration also raises a number of difficult ethical questions that demand careful study. Data privacy, informed consent, transparency, bias reduction, and societal effect are among the issues that are brought to the fore. It is crucial to secure people's privacy, which calls for strong data protection techniques like encryption and anonymization to stop breaches and unwanted access. In order to respect individual autonomy, one must gain informed consent, which enables people to make informed choices about the sharing of their data.

Transparency is essential because it encourages trust and accountability. Organizations must be transparent about how they will use the data they collect. Fundamental bias in past data might sustain unfair results; it is essential to address and mitigate this bias in algorithms to ensure fairness. In order to reduce unfavorable effects, larger societal implications—from digital divisions to employment displacement—require ethical considerations and aggressive action. Organizations may exploit the revolutionary power of IoT and Big Data while keeping ethical standards by developing a culture of ethical technology adoption and ongoing ethical reflection.

13.10 EXPLORING FUTURE HORIZONS

Exciting future possibilities are emerging as IoT and Big Data technologies continue to develop. These two disciplines' mutually beneficial relationship has the ability to change industry and improve our daily lives. Future applications promise to be more complex, enabling proactive decision-making in industries like healthcare, transportation, and energy through real-time data insights. Rapid response times will be made possible by edge computing, which accelerates data processing at the source. A data-driven revolution will be sparked by the growth of IoT to include smart cities, wearable technology, and industrial automation, boosting efficiency and sustainability. Ethics will continue to be the driving force behind conversations about responsible data usage, privacy, and equal access. Additionally, the advent of 5G networks will amplify data connectivity, opening doors to seamless communication among devices. In essence, the future of IoT and Big Data integration is marked by boundless innovation, offering the potential to transform the way we interact with technology, society, and the world around us.

13.11 CONCLUSION

A critical turning point in the digital transformation of societies and industries is being marked by the convergence of IoT and Big Data technologies. Through this connection, hitherto unheard-of possibilities for real-time insights, predictive analytics, and improved decision-making have emerged. It also has a number of difficulties and ethical issues, though, which call for careful navigation. This synthesis has demonstrated the breadth of IoT and Big Data integration, from enhancing maintenance practices through IoT-driven predictive maintenance to protecting privacy and tackling bias in data analysis. Organizations must give data security, transparency, and responsible innovation top priority as we embrace the future. By fostering collaboration, bridging skill gaps, and adhering to ethical principles, we can unlock the transformative potential of this integration while ensuring a sustainable, equitable, and ethically sound future for technology-driven advancements.

REFERENCES

1. Abomhara, M., & Køien, G. M. (2014, May). Security and privacy in the Internet of Things: Status and open issues. In *2014 international conference on privacy and security in mobile systems (PRISMS)* (pp. 1–8). IEEE.

2. Cha, S. C., Hsu, T. Y., Xiang, Y., & Yeh, K. H. (2018). Privacy enhancing technologies in the Internet of Things: Perspectives and challenges. *IEEE Internet of Things Journal, 6*(2), 2159–2187.

3. Paolone, G., Iachetti, D., Paesani, R., Pilotti, F., Marinelli, M., & Di Felice, P. (2022). A holistic overview of the Internet of Things ecosystem. *IoT, 3*(4), 398–434.

4. Ziegeldorf, J. H., Morchon, O. G., & Wehrle, K. (2014). Privacy in the Internet of Things: threats and challenges. *Security and Communication Networks, 7*(12), 2728–2742.

5. Hameed, S., Khan, F. I., & Hameed, B. (2019). Understanding security requirements and challenges in Internet of Things (IoT): A review. *Journal of Computer Networks and Communications, 2019*(1), 9629381.

6. Lee, C. K. M., Cao, Y., & Ng, K. H. (2017). Big data analytics for predictive maintenance strategies. In *Supply Chain Management in the Big Data Era* (pp. 50–74). IGI Global.

7. Biswas, A. R., & Giaffreda, R. (2014, March). IoT and cloud convergence: Opportunities and challenges. In *2014 IEEE World Forum on Internet of Things (WF-IoT)* (pp. 375–376). IEEE.

8. Torre, D., Chennamaneni, A., & Rodriguez, A. (2023). Privacy-preservation techniques for IoT devices: a systematic mapping study. *IEEE Access, 11*, 16323–16345.

9. Rai, A. K., Pokhariya, H. S., Tiwari, K., Vani, V. D., Kumar, D., Kumar, A., & Rana, A. (2023, September). IOT Driven Predictive Maintenance Using Machine Learning Algorithms. In *2023 6th International Conference on Contemporary Computing and Informatics (IC3I)* (Vol. 6, pp. 2674–2678). IEEE.

10. Smith, S. W. (2020). Securing the Internet of Things: An ongoing challenge. *Computer, 53*(6), 62–66.

11. Husnoo, M. A., Anwar, A., Chakrabortty, R. K., Doss, R., & Ryan, M. J. (2021). Differential privacy for IoT-enabled critical infrastructure: A comprehensive survey. *IEEE Access, 9*, 153276–153304.

12. Rana, S. K., Rana, S. K., Rana, A. K., & Islam, S. M. (2022, November). A Blockchain supported model for secure exchange of land ownership: an innovative approach. In *2022 International Conference on Computing, Communication, and Intelligent Systems (ICCCIS)* (pp. 484–489). IEEE.

13. Moleda, M., Małysiak-Mrozek, B., Ding, W., Sunderam, V., & Mrozek, D. (2023). From corrective to predictive maintenance—A review of maintenance approaches for the power industry. *Sensors, 23*(13), 5970.

14. Li, S., Zhao, S., Min, G., Qi, L., & Liu, G. (2021). Lightweight privacy-preserving scheme using homomorphic encryption in industrial Internet of Things. *IEEE Internet of Things Journal, 9*(16), 14542–14550.

14 Application of Artificial Intelligence in the Diagnosis and Treatment of Cancer
A Comprehensive Review

Anindya Nag, Biva Das, Riya Sill,
Anupam Kumar Bairagi, Alok Dutta,
and Pranto Bosu

14.1 INTRODUCTION

The term Artificial Intelligence (AI) pertains to the theoretical ability of robots to perform cognitive functions that are commonly associated with human brains, including but not limited to learning and problem-solving [1]. Individuals who exhibit a proclivity toward observation, wherein they assimilate data from their environment and exhibit appropriate conduct, are more likely to achieve success via AI [2]. AI is an interdisciplinary field that encompasses multiple disciplines such as mathematics, sociology, information systems, communication, cognitive science, neuroscience, and artificial philosophy [3]. AI encompasses various subfields including deep learning (DL), machine learning (ML), and pattern recognition. The acquisition of knowledge and expertise by a machine through experience, rather than manual instruction, is made possible by AI [4]. AI pertains to the cognitive abilities exhibited by computer systems, in contrast to the intellectual capacities demonstrated by animals and humans. The field of AI is concerned with the investigation of intelligent agents, which are systems capable of acquiring information about their surroundings and implementing actions to enhance the likelihood of achieving their objectives [5]. The recent advancements in AI, ML, and DL have had a profound impact on contemporary society and the lifestyles of individuals. AI is a broad field that encompasses a variety of technologies, including sophisticated algorithms, ML, and DL. It is anticipated that AI, ML, and DL will offer automated tools for the timely prognosis, diagnosis, and management of chronic ailments [6]. The diagram of AI is illustrated in Figure 14.1 [7].

Chronic diseases represent a significant challenge within the healthcare domain. Chronic illnesses have been identified as the leading cause of mortality in the medical literature. The cost of treating such diseases has been reported to potentially

DOI: 10.1201/9781032673479-18

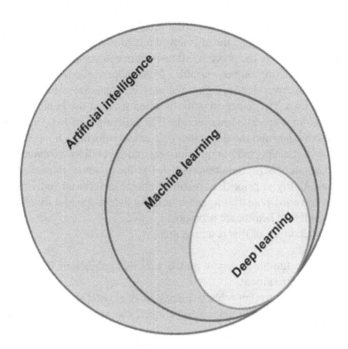

FIGURE 14.1   A schematic representation of AI.

consume up to 70% of a patient's income [8]. Thus, it is imperative to mitigate the mortality risk of the patient. The advancement of medical science has facilitated the enhancement of data-gathering procedures within the healthcare industry [9]. The healthcare dataset comprises the patient's demographic details, medical examination findings, and medical history [10]. The prevalence and incidence of diseases may differ based on the geographical location and living conditions of a particular region. The collection of data pertaining to the patient's living environment and environmental state should be incorporated in addition to illness information.

The healthcare sector has witnessed significant advancements in recent times, primarily attributed to the adoption of information technology (IT). The objective of incorporating IT into the healthcare sector is to enhance the economic and comfortable aspects of an individual's life, similar to the way in which cell phones have simplified life [11]. The feasibility of this proposition could potentially be realized through the implementation of intelligent healthcare systems, including the advancement of intelligent ambulance technology and other comparable hospital services. The study revealed that there existed no statistically significant disparity between male and female patients suffering from chronic ailments within a particular region [12]. Additionally, a considerable number of individuals were admitted to hospitals annually for the management of chronic illnesses. Incorporating both structured and unstructured data yields more precise outcomes compared to solely relying on structured data [13].

Within the medical domain, AI pertains to the methodology of utilizing automated techniques for the purpose of diagnosing and treating patients. The growing utilization of AI in the field of medicine is expected to facilitate the mechanization of numerous professions, thereby affording medical practitioners additional time to undertake tasks that are beyond the purview of machines [14,15]. In the field of ML, it is customary to categorize the process into two main categories: supervised learning, which involves predicting output variables from input variables, and unsupervised learning, which involves identifying output variables that are not directly predicted from input variables. This categorization is often used in the context of clustering different groups for a specific intervention. The utilization of ML in the identification of intricate models and the extraction of medical information has resulted in the exposure of novel concepts to both practitioners and specialists, as stated in reference [16]. The utilization of ML prediction models has the potential to assist clinicians in enhancing their ability to prioritize treatment choice criteria for individual patients. These devices have the potential to autonomously detect specific illnesses in accordance with established healthcare protocols.

The key contributions of this research are:

- Conducting a literature survey on the various applications of AI in cancer diagnosis and treatment.
- After conducting a comprehensive analysis of several approaches employed in the diagnosis and treatment of cancer, the most effective method is determined.

The authors have designed the chapter in such a way that Section 14.2 includes the relevant works and principles, Section 14.3 gives a comparative analysis of different methods, and Section 14.4 outlines the conclusion and further work.

14.1.1 THE ROLE OF AI IN HEALTHCARE

Scientific research indicates that the collaboration between healthcare and technology results in increased benefits for patients. Developing innovative and reliable healthcare solutions utilizing state-of-the-art technology represents a significant accomplishment. AI is being utilized more frequently in the healthcare industry as a diagnostic tool, treatment planner, and physician's assistant [17]. The field of computer vision finds extensive application in various domains such as medical imaging. The system's capability to identify subtle visual cues can potentially enhance the diagnosis of diseases like cavity, cancer, and others. Moreover, research indicates that DL algorithms have been applied in the domain of medical imaging [18]. According to a study conducted by researchers, it was found that deep neural networks exhibited superior performance in comparison to convolutional neural networks and multilayered classifiers when it came to the diagnosis of lung cancer. The present study centers on the application of DL/ML techniques to detect tumors and differentiate them as either malignant or benign, with the aim of facilitating subsequent medical interventions. It is worth noting that comparable machine learning approaches are employed in the realm of cancer diagnosis [19].

The healthcare industry has exhibited a prolonged reluctance to adopt technological advancements, thereby maintaining a predominantly human-centric approach. Historically, robots have been deficient in their ability to comprehend, exhibit finesse, achieve precision, and demonstrate proficiency, in comparison to medical professionals [20]. The initial phases of cancer are challenging to detect, and recurrence frequently occurs subsequent to cessation of treatment. Making dependable prognostications regarding the trajectory of an ailment is also a challenging task. The identification of specific types of cancer at an early stage may present difficulties owing to indistinct symptoms and ambiguous indicators on mammograms and scans. The development of improved prediction models utilizing multivariate data and high-resolution diagnostic technology is a crucial aspect of clinical cancer research [21]. The proliferation of research papers on cancer analysis has been attributed to the utilization of AI methodologies and extensive datasets comprising of past clinical cases to train AI models. The emergence of AI and ML has led to the recognition that machines can possess a degree of intelligence and be utilized for diagnostic purposes [22]. The diagram depicted in Figure 14.2 illustrates the implementation of AI technology within the healthcare industry [23].

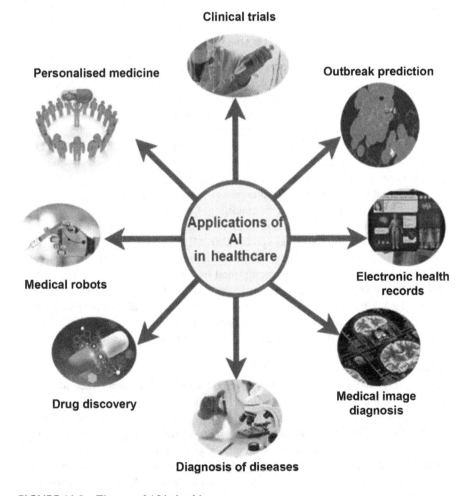

FIGURE 14.2   The use of AI in healthcare.

AI is frequently employed in medical research, primarily to expand the range of applications for current medications and enhance the efficacy of those that are already available [24]. In order to safeguard patient safety and ensure the accuracy of generated data, scholarly investigations have been carried out utilizing AI models for the purpose of detecting instances of misrepresented adverse events [25]. The utilization of AI has demonstrated its significance in aiding medical personnel in alleviating their workload. This is equally significant as its contribution toward prevention, diagnosis, and treatment. The utilization of chatbots in establishing prompt communication with patients and acquiring additional information regarding their symptoms has demonstrated significant utility. Various AI-based predictive models are employed to gain a comprehensive understanding of patients and their risk levels in order to differentiate those who require immediate medical attention from those who do not. The heightened precision of robotic-assisted surgeries has resulted in a surge in demand for such procedures. A number of AI technologies are utilized by surgeons to analyze patient data in real-time during critical surgical procedures [26].

14.1.2 DISEASE DIAGNOSIS AND TREATMENT USING AI

There has been an increasing interest in the utilization of AI and ML within the healthcare industry for the purpose of disease identification and treatment since the 1970s. The fundamental concept underlying ML algorithms is that of computational statistics. Upon submission of the data to the algorithm, it undergoes a learning process wherein it examines and investigates the interrelationships among the data points. The algorithm endeavors to reveal complex patterns between the data that may not be readily comprehensible to a human observer [27]. This characteristic enables its application in the timely identification of cancers, which are often challenging to diagnose but can result in fatal outcomes if not managed effectively. Cancer is a prominent illustration of this type of condition. Distinguishing between early-stage cancerous or tumorous growths and their more malignant counterparts is a challenging task for the human visual system [28]. ML has the capability to accurately detect these initial warning signals. The integration of genome-based tumor sequencing with cognitive computing, as exemplified by IBM's Watson Genomics, has the potential to expedite the process of diagnosis [29]. The researchers at MD Anderson developed a ML technique to forecast acute toxicities in patients with head-and-neck cancer who are receiving radiation therapy [30]. Berg, a prominent pharmaceutical company, has incorporated AI principles in the field of cancer treatment [31].

14.1.2.1 Using AI to Predict Cancer

The etymology of the term cancer can be traced back to the ancient Greek word hapkido, which connotes both crab and tumor. The medical community was first introduced to cancer in the 1600s. This disease is distinguished by the anomalous growth of cells that have the potential to invade or metastasize to other organs. Metastasis in cancer pertains to the unrestrained cellular proliferation that originates from a specific anatomical site and disseminates to other regions of the body. Cancer cells can be classified into two categories: benign and malignant. Malignant cells exhibit metastatic behavior and are considered to be more deleterious than benign cells, which lack the ability to disseminate to distant sites. As a result of the disease's elevated mortality and recurrence rates, its treatment is both resource-intensive and costly [32]. The acquisition of an accurate initial diagnosis is of utmost importance in enhancing the likelihood of survival among individuals afflicted with cancer. Inherited disorders such as the aforementioned are attributed to genetic aberrations that impact the regulation of fundamental cellular processes, such as proliferation and differentiation. Additional alterations will manifest as the proliferation of neoplastic cells advances. The prognostic assessments for cancer patients encompass the likelihood of disease metastasis, recurrence frequency, and post-treatment survival duration. The probabilities of both a novel cancer diagnosis and the recurrence of a prior cancer are computed. Subsequently, it is imperative to generate forecasts regarding the progression of the ailment, the longevity of the individual, the receptiveness of the neoplasm to therapeutic interventions, and the patient's overall condition.

The initial phases of cancer are difficult to detect, and there is a tendency for it to recur following cessation of treatment. Making accurate prognoses regarding the progression of a medical condition is also a challenging task. The identification of certain types of cancer in their early stages may present difficulties owing to ambiguous symptoms and perplexing indicators on mammograms and scans. The development of improved prediction models utilizing multivariate data and high-resolution diagnostic technology is imperative in the realm of clinical cancer research. A preliminary examination of scholarly literature indicates that the utilization of AI techniques in conjunction with extensive data sets comprising past clinical studies for the purpose of constructing AI models has resulted in a noteworthy surge in the quantity of publications pertaining to cancer analysis. The research findings suggest that traditional analytical methods, such as multivariate data analysis, exhibit lower levels of precision compared to AI techniques [33]. The amalgamation of AI and advanced bioinformatics technologies presents a promising avenue for significantly improving the accuracy of medical diagnosis, prognosis, and prediction. ML is increasingly prevalent and is considered to be a more precise designation. ML is a subfield of AI that involves the development of algorithms and statistical models that enable computer systems to automatically improve their performance on a specific task by learning from data without being explicitly programmed. The primary objective of ML is to create predictive models that can be used to make accurate predictions or decisions based on input data. The possibility exists for these algorithms to evaluate a patient's probability of survival through the acquisition of logical patterns derived from extensive amounts of past data. ML has been utilized in various domains, including the improvement of prognostic outcomes. The ability to offer a prognosis is an essential therapeutic competency, especially for healthcare professionals who are involved in the care of individuals with cancer. The accuracy of early cancer diagnosis and prognosis is enhanced by ML techniques, which have demonstrated efficacy in predicting cancer susceptibility, recurrence, and survival. ML possesses the capability to enhance clinical outcomes in the context of patient care. Researchers in the domains of medicine and computer science have been endeavoring to enhance ML techniques to classify cancer patients into high- and low-risk recurrence groups, thereby enabling more precise prognostic care. Alternatively, AI could potentially be employed to evaluate and analyze multifaceted data from diverse patient examinations, thereby enhancing the precision of cancer prognosis, disease progression, and survival time prediction [34].

14.1.2.2 Difficulties in Cancer Diagnosis Using AI

Research has demonstrated that AI is capable of proficiently handling non-linear relationships, high availability, massively parallel computation, learning, and other related tasks. Due to its autonomous adaptability, this technology has the capability to analyze both quantitative and qualitative data concurrently, rendering it advantageous in numerous clinical trials. Undoubtedly, AI holds numerous potential applications within the medical domain. Furthermore, by leveraging the various facets of clinical heterogeneity, it serves to address the prevailing dearth of objectivity and inclusivity in contemporary expert systems [35]. The implementation of AI in medical education could potentially provide advantages for medical facilities by enhancing the teaching of diagnostic and decision-making skills to medical students and residents. Within a burgeoning corpus of scholarly literature, researchers extol the enhanced diagnostic and prognostic capacities of machine learning-driven computational systems. DL techniques are having an impact on the way in which imaging data is analyzed and interpreted. In comparison to radiologists, the aforementioned outcomes have the potential to enhance sensitivity and reduce instances of false positives. Nonetheless, there exists a potential hazard of overfitting the training data, which could result in inferior performance under specific circumstances. According to literature, it has been observed that on several occasions, the outcomes obtained from highly precise models such as boosted trees, random forests, and neural networks tend to be unintelligible. Conversely, the outcomes derived from more comprehensible models such as logistic regression, Naive Bayes, and single decision trees are comparatively less precise [36].

14.2 LITERATURE REVIEW

A number of scholars explicate their discoveries as illustrated below.

- **Deep Learning (DL)**

  Noia et al. [37] revealed in this study that AI tools are increasingly being utilized to identify optimal brain tumor survival assessment and prediction techniques, in response to the rising therapeutic requirements. The growth has been expedited by the utilization of computational resources and publicly available databases. The present study provides a narrative assessment of the utilization of AI in predicting the survival rates of brain tumor patients, with a specific emphasis on Magnetic Resonance Imaging (MRI). Extensive searches were conducted on PubMed and Google Scholar utilizing Boolean research queries based on MeSH keywords, with the search parameters narrowed to the years 2012–2022. Fifty scholarly articles pertaining to the topics of ML, DL, radio-mics, and survival evaluation utilizing conventional imaging were selected. The performance of computational approaches varies depending on the task at hand, and the most suitable approach is contingent upon a multitude of factors. The findings indicate that the utilization of quantitative imaging properties can prove advantageous in AI applications, such as predicting the survival rates of cancer patients.

- **Lung Nodule Analysis (LUNA-16)**

  Wang et al. [38] said that lung cancer poses a significant threat to human health and is considered one of the most perilous forms of cancer. The effective management of lung cancer necessitates a comprehensive identification of the pathogenic subtype through diagnostic procedures. Historically, the identification of the pathological manifestation of lung cancer has required a histological examination that is both invasive and lengthy in duration. The present research introduces an innovative residual neural network model aimed at detecting the pathological subtype of lung cancer through the analysis of Computed Tomography (CT) images. This study aimed to explore the medical-to-medical transfer learning technique as a potential solution to the limited availability of CT images in clinical practice. The present study involves the utilization of a pre-trained residual neural network that has been trained on the publicly available medical images dataset LUNA-16, which is specifically designed for the analysis of lung nodules. Subsequently, the network is fine-tuned using a proprietary dataset of lung cancer images that has been collected from the Shandong Provincial Hospital. The results of data experiments indicate that the proposed strategy exhibits superior performance compared to currently available systems that were trained using 2054 labels. The proposed strategy achieves an accuracy of 85.71% in identifying pathological types of lung cancer from CT scans obtained from LUNA-16 dataset. The findings indicate that this methodology surpasses the performance of AlexNet, VGG16, and DenseNet, offering a noninvasive and efficient means of detecting cancer for diagnostic purposes.

- **Convolutional Neural Network (CNN)**

  Haq et al. [39] mentioned that the classification of brain tumors is a critical aspect of medical practice, and the utilization of the Internet of Things (IoT) has played a significant role in the detection of brain cancer. The timely detection of brain cancer is a significant medical concern, and numerous researchers have developed diagnostic methodologies. An automated diagnostic system was proposed by researchers to assist physicians in the diagnosis and treatment of brain cancer. The proposed integrated framework initially integrated a CNN model for the purpose of extracting deep features from MRI. The classification of retrieved characteristics for cancer prediction is accomplished by a Long Short-Term Memory model. The utilization of augmentation techniques resulted in an increase in the volume of data, leading to an enhancement in the efficacy of the model. The methodology employed in this study involved the utilization of hold-out cross-validation for both training and validation of the approach. The utilization of multiple measures to assess the model can aid in the prediction of cancer. The experimental results indicate that the proposed diagnostic framework, which combines CNN and Long Short-Term Memory (LSTM) techniques, exhibited superior performance compared to the baseline solutions commonly used in the IoT industry. The superior performance of the model may be attributed to various factors such as data preprocessing, model parameters including layer count, optimizer, and activation functions, as well as data augmentation techniques. The findings indicate that the CNN-LSTM methodology exhibits strong performance and is suggested as a suitable option for the IoT framework in the context of detecting brain cancer.

Heuvelmans et al. [40] reported the utilization of an independent dataset of indeterminate nodules from a European multicenter experiment to retrospectively validate the Lung Cancer Prediction CNN (LCP-CNN). The LCP-CNN was initially trained on data from the United States' screening program, enabling it to classify cancerous nodules while excluding benign nodules and maintaining sensitivity for lung cancer. The LCP-CNN algorithm was trained using CT data from the United States in order to assign a malignancy score to individual nodules. The LUCINDA study recruited patients from three tertiary referral centers located in the United Kingdom, Germany, and the Netherlands. The study employed CT images, which consisted of 2106 nodules, including 205 lung malignancies, for validation purposes. The objective is to establish boundaries for the morbidity rating, with a sensitivity level of at least 99.9%, utilizing data from the National Lung Screening Trial, as mandated by the pre-existing benign nodule rule-out test. The findings indicate that the Area-Under-the-ROC Curve (AUC) was 94.5% across the European centers. The study found that the sensitivity of the diagnostic test was 99.0%, indicating that a significant proportion of nodules (22.1%) were benign. This finding has important implications for clinical practice, as it allows a substantial proportion of patients (18.5%) to avoid unnecessary follow-up scans. Furthermore, this diagnostic tool provides a reliable means of predicting lung cancer.

Zadeh et al. [41] suggested that brain tumor histology-stained images were utilized for survival categorization. This study employs The Cancer Genome Atlas (TCGA) and Adelaide samples to evaluate the efficacy of Deep CNN (DCNN) algorithms. The DeepSurvNet classifier was trained on brain cancer samples from TCGA by GoogleNet. The DeepSurvNet model achieved a 99.99% accuracy rate in predicting the time to occurrence of brain tumors based on classified patches. Histological examinations of full-slide hematoxylin and eosin-stained biopsies have the potential to reveal the presence of malignant pathology and its associated clinical ramifications. While it would be advantageous to strategize treatment and allocate time for preclinical research to administer personalized therapies, presently, there are no automated algorithms of high reliability that establish a correlation between histopathological images and the survival of brain cancer patients. The DeepSurvNet algorithm is utilized to classify the prognoses of brain cancer patients into four distinct categories, which are determined based on the histopathological images. These categories are identified as follows: Class I, with a predicted survival time of 0–6 months; Class II, with a predicted survival time of 6–12 months; Class III, with a predicted survival time of 12–24 months; and Class IV, with a predicted survival time exceeding 24 months. The authors assessed the performance of DeepSurvNet on novel samples following its training on TCGA via the employment of the DCNN methodology. The dataset utilized in this study was a publicly available collection of brain cancer data. The DeepSurvNet model has demonstrated high accuracy in predicting brain cancer prognoses through the analysis of histological images, achieving accuracies of 0.99 and 0.8 on the respective datasets. The study on mutation frequency ultimately revealed mutation rates and types specific to each class, thereby corroborating the notion of a unique genetic signature associated with patient survival. The DeepSurvNet model has demonstrated the ability to make predictions regarding the survival of patients with brain cancer.

Tandel et al. [42] found that neoplasms of the central nervous system, specifically the brain, constitute 10% of all cancer-related mortalities in both genders. Brain tumors are observed to progress in 40% of cases of lung and breast cancer due to metastasis. The rise in the incidence of brain tumors underscores the necessity for a prompt, dependable, and non-intrusive computer-assisted diagnostic mechanism. The utilization of Convolutional Networks within the AI framework for transmission purposes has resulted in an enhancement of brain tumor grade classification derived from MRI data, as evidenced by the CCN model. The superiority of CNN-based DL over six ML models has been demonstrated on five separate multiclass tumor datasets. The study conducted three cross-validation procedures, namely K2, K5, and K10, to evaluate the performance of a transfer learning program based on a CNN using the AlexNet architecture. The overall accuracy rate of the program was found to be 93.74%. The corresponding accuracy values for K2, K5, and K10 were 95.97, 96.65, and 87.14, respectively. The tumor separation index was utilized to validate the optimal model based on synthetic data consisting of eight classes. The domain of multiclass brain tumor grading is one in which AI systems based on transfer learning demonstrate superior performance compared to those based on traditional ML methods. The findings demonstrate the effectiveness of an AI system based on transfer learning in the classification of brain tumors into multiple grades, with the system exhibiting superior performance compared to ML techniques.

Ellah et al. [43] suggested that there has been a rise in the incidence of brain tumor's among individuals. The process of identifying, locating, and classifying tumors through MRI is a challenging and imprecise task. The present research offers a methodological approach utilizing multiple models for the automated detection and localization of brain tumors. The approach is comprised of two distinct stages. During the preliminary stage of the system, data cleaning and feature extraction will be conducted through the utilization of a CNN. Subsequently, feature classification will be executed by means of an Error-Correcting Output Codes Support Vector Machine (ECOC-SVM). The initial phase of brain tumor identification involves the categorization of MRI scans into two groups, namely normal and abnormal. The second component of the system utilizes a CNN with five layers that is region-based in order to accurately identify the location of the tumor within the MRIs that have been distorted. The efficacy of the initial stage was assessed through the utilization of three unique CNN models, namely AlexNet, VGG-16, and VGG-19. The highest detection accuracy achieved by AlexNet was 99.55%, as determined by analyzing 349 images from the Reference Image Collection to Evaluate Response (RIDER) Brain MRI database. The utilization of the Brain Tumor Segmentation (BraTS) 2013 dataset is employed to achieve a DICE score of 0.87 in the identification of brain tumors through the use of 3D MRIs. The findings of an empirical inquiry indicate that the proposed approach for tumor detection exhibited superior performance compared to non-DL systems that were designed for facilitating cancer prognosis. The present research offers empirical support for the proposed approach in improving the identification and pinpointing of cancer.

- **Random Forest (RF) and Adaptive Boosting (AdaBoost)**

  Johnson et al. [44] proposed a framework consisting of multiple stages for the creation of an AI-driven decision-support system capable of predicting the survival of lung cancer patients after a period of five years. The dataset obtained from the National Institute of Health, namely the Surveillance, Epidemiologic, and End Results dataset spanning from 1973 to 2015, enables us to perform an analysis that assesses the proposed methodology. The initial stage involves the establishment of a specific objective and undertaking any essential data preprocessing. During the second phase, six unique AI algorithms are employed, whereby particle swarm optimization is employed for the purpose of feature selection, and cross-validation is utilized for hyperparameter tuning. The aforementioned domain encompasses a variety of methodologies, including but not limited to Logistic Regression, Decision Trees (DT), RF, AdaBoost, Artificial Neural Network (ANN), and Naive Bayes. The findings indicate that the RF and AdaBoost models exhibit superior performance compared to the other models, with an AUC rate of 0.94 attributed to their specific application.

- **Probabilistic Neural Network (PNN)**

  Sannasi Chakravarthy et al. [45] explained that there is a higher incidence of lung cancer diagnosis in men compared to women. Additionally, lung cancer is ranked as the third most prevalent type of cancer in women. The timely identification of lung cancer has been observed to result in a reduction in mortality rates on a global scale. Medical imaging is necessary for the early detection of lung cancer, prior to the manifestation of symptoms. The present study proposes the utilization of an automated classification system for the early detection of lung cancer. The present study involves the integration of CT imaging of the lungs with PNN classification. Following the preprocessing of the lung images, feature extraction and selection are performed using the Gray-Level Co-Occurrence Matrix and the Chaotic Crow Search Algorithm (CCSA), respectively. The evaluation of lung cancer involves the presentation of selectivity, sensitivity, negative predictive value, and positive predictive value. According to the results, the utilization of CCSA-based feature selection could potentially lead to precise identification of lung cancer, with the possibility of attaining a 90%

accuracy rate. The research results indicate that the PNN model, utilizing feature selection based on CCSA, demonstrates superior performance in accurately predicting lung cancer levels.

- **Elastic Net Classifiers (ENC)**

Barker et al. [46] described that the utilization of computerized pathology image analysis has the potential to enhance automated diagnosis and the discovery of disease subtypes. The size of digital pathology images poses a challenge for computer processing, while the presence of tissue sections that are not directly related to the illness may potentially lead to erroneous results in automated diagnosis algorithms. Examining the specific attributes of sick photographs in a gradual and systematic manner can aid in mitigating associated challenges. Initially, it is recommended to examine the overall composition of the slide, taking into account its general layout and segmented areas, in order to ascertain its form, chromaticity, and surface characteristics. The technique of dimensionality reduction is employed to condense feature dimensions and generate representative subsets that can be utilized to quantify the diversity of images in tiled regions. The second step involves the evaluation of the group representation of a sample tile. The ENC technique is employed for the purpose of cancer prediction. Weighted voting is a method of tallying votes that assigns different values to each vote based on certain criteria. This approach can be used to analyze the results of a survey or election and determine the outcome of a particular issue or decision. In the context of diagnosing a slide, weighted voting totals may be used to assess the relative importance of different factors or variables that could be contributing to the problem. The method was able to accurately distinguish between glioblastoma multiforme ($N = 182$) and lower-grade glioma ($N = 120$) with a prediction accuracy of 95.5% in the context of brain cancer patients. Following a fivefold cross-validation procedure, the aforementioned approach successfully classified all datasets in the MICCAI Pathology Classification Challenge, achieving a 100% classification rate. The method exhibited robustness and consistency across various cancer diagnoses with a high degree of accuracy. According to the findings, ENC has the ability to accurately distinguish between lung and breast cancer through mechanical means.

- **Artificial Neural Network (ANN)**

Shukla et al. [47] found that lung cancer is promoted by smoking, radiation, and chemicals, resulting in chronic tension and dysfunction of the autonomic nervous system. The non-linear analysis of heart rate variability (HRV) has the potential to differentiate between individuals with lung cancer and those who are healthy. HRV indicators were obtained by analyzing the electrocardiograms of both lung cancer patients and individuals in good health. The HRV values of the Eastern Cooperative Oncology Group exhibited a decrease from 1 to 4. The HRV of males exhibited greater magnitude than that of females. Moreover, the presence of a did not exert any influence on HRV, as n non-linear HRV features were found to be pivotal. The results indicate that ANN achieved a score of 83.3%, while ECOG2, ECOG3, and ECOG4 received scores of 50%, 90%, and 95%, respectively. The control group achieved a score of 86.7%. The findings indicate that the utilization of an ANN with non-linear HRV values as input for prediction purposes yielded a score of 93.09% and attained output accuracy, thereby exhibiting its potential prognostic significance.

- **Recurrent Neural Network (RNN)**

    Selvanambi et al. [48] mentioned that challenges in therapy are a common occurrence for individuals. Cancer is considered to be one of the most catastrophic diseases. The difficulty of early detection is compounded by the delayed onset of negative consequences. The prognostication of lung cancer is a crucial aspect of the analytical approach as it enhances the efficacy of treatment interventions. Enhancing a patient's chances of survival is a challenging strategy to implement. The present study employs the Levenberg-Marquardt model and the glowworm swarm optimization approach to introduce an enhanced neural network system, namely the RNN technique, for effectively handling multimodal disease data. The efficacy of the proposed techniques was evaluated using data and a benchmark dataset. The outcomes demonstrate that higher-order recurrent neural networks, which were optimized using conventional neural networks, achieved cancer prediction accuracy up to 98%.

    Table 14.1 shows a diverse variety of authors who employed the technique and presented their findings.

TABLE 14.1
Summarize the Reviewed Literature

Author Name	Methods Used	Outcome
Noia et al. [37]	DL	The research findings indicate the significance of utilizing quantitative imaging characteristics in AI applications, such as forecasting cancer survival rates
Wang et al. [38]	LUNA-16	The mentioned approach presents a highly efficacious means of detecting cancer for diagnostic purposes, without the need for invasive procedures. Furthermore, it has demonstrated superior performance compared to AlexNet, VGG16, and DenseNet
Haq et al. [39]	CNN-LSTM	The results indicate that the CNN-LSTM approach is a feasible alternative for detecting brain cancer within an IoT framework
Heuvelmans et al. [40]	LCP-CNN	Upon analyzing the collective European canters, the data indicates that the Area-Under-the-ROC Curve yielded a result of 94.5%. The predictive accuracy of lung cancer was found to be 99.0%, resulting in the identification of 22.1% of nodules with high precision. This allowed 18.5% of patients to avoid additional scans
Zadeh et al. [41]	DCNN	The findings lend support to the conjecture that a distinct genetic profile is linked to a patient's likelihood of survival and demonstrate that the frequency and categories of mutations differ across categories. DeepSurvNet can be utilized to predict the survival outcomes of patients with brain cancer

(Continued)

TABLE 14.1 (*Continued*)
Summarize the Reviewed Literature

Author Name	Methods Used	Outcome
Tandel et al. [42]	CNN	The results indicate that a transfer learning approach utilizing AI technology outperforms ML techniques in the classification of brain tumors with multiple classes
Ellah et al. [43]	CNN	The results indicate that the suggested methodology for identifying tumor's outperformed non-deep learning models that were fine-tuned for cancer prediction. The results of this study indicate that the implementation of the methodology enhances the ability to detect and localize cancer
Johnson et al. [44]	RF and AdaBoost	In comparison to alternative models, it can be observed that the AUC for RF and AdaBoost is 0.94. This indicates that the practical application of these models produces superior outcomes
Sannasi Chakravarthy et al. [45]	PNN	The study reveals that the PNN model, which employs CCSA-based feature selection, exhibits superior performance in assessing the precision of lung cancer prediction
Barker et al. [46]	ENC	The findings indicate that the utilization of ENC exhibits potential for the automated identification of lung and breast cancer
Shukla et al. [47]	ANN	The prediction utility of an ANN was demonstrated through the utilization of non-linear heart rate variability (HRV) values as input, resulting in an output accuracy of 93.09%
Selvanambi et al. [48]	RNN	Based on the results, it has been observed that neural networks that are adjusted with a higher degree of recurrence have the ability to attain an accuracy rate of up to 98% in predicting cancer

14.3   COMPARATIVE ANALYSIS

The main objective of this comparative analysis is to evaluate and contrast various methodologies utilized in the prediction of cancer. This study focuses on the utilization of AI methodologies for the purpose of predicting cancer. A comparative analysis of studies is conducted to predict chronic diseases using accuracy values. The accuracy values are obtained through a comparative analysis of various AI methodologies for cancer prediction, including but not limited to the DCNN, RNN, LUNA-16, PNN, and CNN techniques. According to Wang et al. [44], the LUNA-16 technique yielded a minimum accuracy of 85.71%. Meanwhile, Ellah et al. [47] reported an accuracy value of 99.55% for the prediction of chronic cancer using AI approaches with the CNN method.

Table 14.2 presents a comparative analysis of the accuracy values of various AI methods in predicting cancer. Table 14.2 presents comparative analyses of different models.

Figure 14.3 shows a comparison graph for the accuracy value.

TABLE 14.2

Comparative Study of Reviewed Literature

Author	Technology	Accuracy (%)
Wang et al. [38]	LUNA-16	85.71
Zadeh et al. [41]	DCNN	99.00
Ellah et al. [43]	CNN	99.55
Sannasi Chakravarthy et al. [45]	PNN	90.00
Selvanambi et al. [48]	RNN	98.00

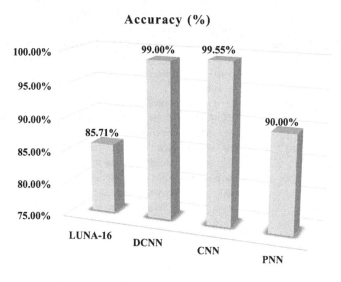

FIGURE 14.3   Accuracy as shown graphically.

14.4 CONCLUSION AND FUTURE SCOPE

This chapter entail doing a comprehensive review of the existing literature pertaining to the diverse uses of AI in the domain of cancer detection and treatment. Upon conducting an extensive analysis of several ways utilized in the detection and treatment of cancer, the most efficacious strategy has been ascertained. The CNN methodology is utilized to enhance the precision metric in forecasting cancer. Following the evaluation of various methodologies, the comparative analysis indicates that the CNN approach, a specific domain of brain cancer prognosis, exhibits a precision rate of 99.55% when subjected to radiation therapy immediately subsequent to tumors detection. The success rate of the LUNA-16 mission was 85.71%. The DCNN achieved an accuracy rate of 99.00%. The PNN achieved a score of 90.00%. The RNN achieved a classification accuracy of 98.00%. The application of AI is gradually transforming medical diagnosis, representing a future direction and trend in the evolution of healthcare. This study presents an overview of the concept and development of AI in the context of cancer diagnosis, as well as its current applications and potential for future use in both tumors' diagnosis and treatment.

REFERENCES

[1] Nishant, R., Kennedy, M., & Corbett, J. (2020). Artificial intelligence for sustainability: Challenges, opportunities, and a research agenda. *International Journal of Information Management*, 53, 102104. https://doi.org/10.1016/j.ijinfomgt.2020.102104.

[2] Innes, G. (2015). Sorry-we're full! access block and accountability failure in the health care system. *CJEM*, 17(2), 171–179. https://doi.org/10.2310/8000.2014.141390.

[3] Mathur, S., & Modani, U. S. (2016). Smart city - A gateway for artificial intelligence in India. *2016 IEEE Students' Conference on Electrical, Electronics and Computer Science (SCEECS)*. Bhopal, India. https://doi.org/10.1109/sceecs.2016.7509291

[4] Navarathna, P. J., & Malagi, V. P. (2018). Artificial intelligence in smart city analysis. *2018 International Conference on Smart Systems and Inventive Technology (ICSSIT)*. Tirunelveli, India. https://doi.org/10.1109/icssit.2018.8748476.

[5] Gambhir, S., Malik, S. K., & Kumar, Y. (2016). Role of soft computing approaches in healthcare domain: A mini review. *Journal of Medical Systems*, 40(12), 1–15. https://doi.org/10.1007/s10916-016-0651-x.

[6] Srivastava, S., Bisht, A., & Narayan, N. (2017). Safety and security in smart cities using artificial intelligence - A review. *2017 7th International Conference on Cloud Computing, Data Science & Engineering - Confluence*. Noida, India. https://doi.org/10.1109/confluence.2017.7943136.

[7] Miraftabzadeh, S. M., Foiadelli, F., Longo, M., & Pasetti, M. (2019). A survey of machine learning applications for power system analytics. *2019 IEEE International Conference on Environment and Electrical Engineering and 2019 IEEE Industrial and Commercial Power Systems Europe (EEEIC/I & CPS Europe)*. Genova, Italy. https://doi.org/10.1109/eeeic.2019.8783340.

[8] Battineni, G., Sagaro, G. G., Chinatalapudi, N., & Amenta, F. (2020). Applications of machine learning predictive models in the chronic disease diagnosis. *Journal of Personalized Medicine*, 10(2), 21. https://doi.org/10.3390/jpm10020021.

[9] Manjulatha, B., & Pabboju, S. (2021). An ensemble model for predicting chronic diseases using machine learning algorithms. *Smart Computing Techniques and Applications*, 2, 337–345. https://doi.org/10.1007/978-981-16-1502-3_34.

[10] Jen, C.-H., Wang, C.-C., Jiang, B. C., Chu, Y.-H., & Chen, M.-S. (2012). Application of classification techniques on development an early-warning system for chronic illnesses. *Expert Systems with Applications*, 39(10), 8852–8858. https://doi.org/10.1016/j.eswa.2012.02.004.

[11] Gupta, D., Khare, S., & Aggarwal, A. (2016). A method to predict diagnostic codes for chronic diseases using machine learning techniques. *2016 International Conference on Computing, Communication and Automation (ICCCA)*. Greater Noida, India. https://doi.org/10.1109/ccaa.2016.7813730.

[12] Chen, M., Hao, Y., Hwang, K., Wang, L., & Wang, L. (2017). Disease prediction by machine learning over big data from healthcare communities. *IEEE Access*, 5, 8869–8879. https://doi.org/10.1109/access.2017.2694446.

[13] Ge, R., Zhang, R., & Wang, P. (2020). Prediction of chronic diseases with multi-label neural network. *IEEE Access*, 8, 138210–138216. https://doi.org/10.1109/access.2020.3011374.

[14] MacLeod, H., Yang, S., Oakes, K., Connelly, K., & Natarajan, S. (2016). Identifying rare diseases from behavioural data: A machine learning approach. *2016 IEEE First International Conference on Connected Health: Applications, Systems and Engineering Technologies (CHASE)*. Washington, USA. https://doi.org/10.1109/chase.2016.7

[15] Takura, T., Hirano Goto, K., & Honda, A. (2021). Development of a predictive model for integrated medical and long-term care resource consumption based on health behaviour: Application of healthcare big data of patients with circulatory diseases. *BMC Medicine*, 19(1), 1–16. https://doi.org/10.1186/s12916-020-01874-6.

[16] Zufferey, D., Hofer, T., Hennebert, J., Schumacher, M., Ingold, R., & Bromuri, S. (2015). Performance comparison of multi-label learning algorithms on clinical data for chronic diseases. *Computers in Biology and Medicine*, 65, 34–43. https://doi.org/10.1016/j. compbiomed.2015.07.017.

[17] Sunarti, S., Fadzlul Rahman, F., Naufal, M., Risky, M., Febriyanto, K., & Masnina, R. (2021). Artificial intelligence in healthcare: Opportunities and risk for future. *Gaceta Sanitaria*, 35, 67–70. https://doi.org/10.1016/j.gaceta.2020.12.019.

[18] Lee, T. H., Chen, J.-J., Cheng, C.-T., & Chang, C.-H. (2021). Does artificial intelligence make clinical decision better? A review of artificial intelligence and machine learning in acute kidney injury prediction. *Healthcare*, 9(12), 1662. https://doi.org/10.3390/healthcare9121662.

[19] Itahashi, K., Kondo, S., Kubo, T., Fujiwara, Y., Kato, M., Ichikawa, H., Koyama, T., Tokumasu, R., Xu, J., Huettner, C. S., Michelini, V. V., Parida, L., Kohno, T., & Yamamoto, N. (2018). Evaluating clinical genome sequence analysis by Watson for genomics. *Frontiers in Medicine*, 5, 1–10. https://doi.org/10.3389/fmed.2018.00305.

[20] Feroze, S. (2022). Impact of artificial intelligence on professional autonomy of pathologists. *Autonomy of Pathologists* (Doctoral dissertation, Toronto Metropolitan University). https://doi.org/10.32920/19008740.v1.

[21] Kulikowski, C. A., & Weiss, S. M. (2019). Representation of expert knowledge for consultation: The CASNET and expert projects. *Artificial Intelligence in Medicine*, 21–55. https://doi.org/10.4324/9780429052071-2.

[22] Cassidy, J. W. (2020). Applications of machine learning in drug discovery I: Target discovery and small molecule drug design. In John W. Cassidy and Belle Taylor(Eds) *Artificial Intelligence in Oncology Drug Discovery and Development*. IntechOpen, London,United Kingdom, pp. 65–80, https://doi.org/10.5772/intechopen.93159.

[23] Pandya, S., Thakur, A., Saxena, S., Jassal, N., Patel, C., Modi, K., Shah, P., Joshi, R., Gonge, S., Kadam, K., & Kadam, P. (2021). A study of the recent trends of immunology: Key challenges, domains, applications, datasets, and future directions. *Sensors*, 21(23), 7786. https://doi.org/10.3390/s21237786.

[24] Pandya, S., & Ghayvat, H. (2021). Ambient acoustic event assistive framework for identification, detection, and recognition of unknown acoustic events of a residence. *Advanced Engineering Informatics*, 47, 101238. https://doi.org/10.1016/j.aei.2020.101238.

[25] Cheng, F., & Cummings, J. (2022). Artificial intelligence in Alzheimer's drug discovery. In Jeffrey Cummings, Jefferson Kinney and Howard Fillit (Eds) *Alzheimer's Disease Drug Development*, Cambridge University Press, Cambridge, England. pp. 62–72. https://doi.org/10.1017/9781108975759.007.

[26] Lee, D. H., & Yoon, S. N. (2021). Application of artificial intelligence-based technologies in the healthcare industry: Opportunities and challenges. *International Journal of Environmental Research and Public Health*, 18(1), 271. https://doi.org/10.3390/ijerph18010271.

[27] Bagley, A. F., Garden, A. S., Reddy, J. P., Moreno, A. C., Frank, S. J., Rosenthal, D. I., Morrison, W. H., Gunn, G. B., Fuller, C. D., Shah, S. J., Ferrarotto, R., Sturgis, E. M., Gross, N. D., & Phan, J. (2020). Highly conformal reirradiation in patients with prior oropharyngeal radiation: Clinical efficacy and toxicity outcomes. *Head & Neck*, 42(11), 3326–3335. https://doi.org/10.1002/hed.26384.

[28] Bibault, J.-E., Giraud, P., & Burgun, A. (2016). Big data and machine learning in radiation oncology: State of the art and future prospects. *Cancer Letters*, 382(1), 110–117. https://doi.org/10.1016/j.canlet.2016.05.033.

[29] Kyrarini, M., Lygerakis, F., Rajavenkatanarayanan, A., Sevastopoulos, C., Nambiappan, H. R., Chaitanya, K. K., Babu, A. R., Mathew, J., & Makedon, F. (2021). A survey of robots in healthcare. *Technologies*, 9(1), 8. https://doi.org/10.3390/technologies9010008.

[30] Ghayvat, H., Awais, M., Pandya, S., Ren, H., Akbarzadeh, S., Chandra Mukhopadhyay, S., Chen, C., Gope, P., Chouhan, A., & Chen, W. (2019). Smart aging system: Uncovering the hidden wellness parameter for well-being monitoring and anomaly detection. *Sensors*, 19(4), 766. https://doi.org/10.3390/s19040766.

[31] Rong, G., Mendez, A., Bou Assi, E., Zhao, B., & Sawan, M. (2020). Artificial intelligence in healthcare: Review and prediction case studies. *Engineering*, 6(3), 291–301. https://doi.org/10.1016/j.eng.2019.08.015.

[32] Meyer, J., Khademi, A., Têtu, B., Han, W., Nippak, P., & Remisch, D. (2021). Impact of artificial intelligence, with and without information, on pathologists' decisions: An experiment. J Am Med Inform Assoc. 2022 Sep 12;29(10):1688–1695. https://doi.org/10.21203/rs.3.rs-614881/v1.

[33] Simmons, C., McMillan, D. C., Tuck, S., Graham, C., McKeown, A., Bennett, M., O'Neill, C., Wilcock, A., Usborne, C., Fearon, K. C., Fallon, M., & Laird, B. J. (2019). "How long have i got?" - A prospective cohort study comparing validated prognostic factors for use in patients with advanced cancer. *The Oncologist*, 24(9), 960–967. https://doi.org/10.1634/theoncologist.2018-0474

[34] Kourou, K., Exarchos, K. P., Papaloukas, C., Sakaloglou, P., Exarchos, T., & Fotiadis, D. I. (2021). Applied machine learning in cancer research: A systematic review for patient diagnosis, classification and prognosis. *Computational and Structural Biotechnology Journal*, 19, 5546–5555. https://doi.org/10.1016/j.csbj.2021.10.006.

[35] Jin, S., Wang, B., Zhu, Y., Dai, W., Xu, P., Yang, C., Shen, Y., & Ye, D. (2019). Log odds ,and lymph node ratio. *Journal of Cancer*, 10(1), 249–256. https://doi.org/10.7150/jca.27399.

[36] You, R., Liu, Y. P., Lin, M., Huang, P. Y., Tang, L. Q., Zhang, Y. N., Pan, Y., Liu, W. L., Guo, W. B., Zou, X., Zhao, K. M., Kang, T., Liu, L. Z., Lin, A. H., Hong, M. H., Mai, H. Q., Zeng, M. S., & Chen, M. Y. (2019). Relationship of circulating tumor cells and Epstein-barr virus DNA to progression-free survival and overall survival in metastatic nasopharyngeal carcinoma patients. *International Journal of Cancer*, 145(10), 2873–2883. https://doi.org/10.1002/ijc.32380.

[37] di Noia, C., Grist, J. T., Riemer, F., Lyasheva, M., Fabozzi, M., Castelli, M., Lodi, R., Tonon, C., Rundo, L., & Zaccagna, F. (2022). Predicting survival in patients with brain tumors: Current state-of-the-art of AI methods applied to MRI. *Diagnostics*, 12(9), 2125. https://doi.org/10.3390/diagnostics12092125.

[38] Wang, Y., Wu, B., Zhang, N., Liu, J., Ren, F., & Zhao, L. (2020). Research progress of computer aided diagnosis system for pulmonary nodules in CT images. *Journal of X-Ray Science and Technology*, 28(1), 1–16. https://doi.org/10.3233/xst-190581.

[39] Haq, A. U., Li, J. P., Agbley, B. L., Khan, A., Khan, I., Uddin, M. I., & Khan, S. (2022). IIMFCBM: Intelligent integrated model for feature extraction and classification of brain tumors using MRI clinical imaging data in IOT-healthcare. *IEEE Journal of Biomedical and Health Informatics*, 26(10), 5004–5012. https://doi.org/10.1109/jbhi.2022.3171663.

[40] Heuvelmans, M. A., van Ooijen, P. M. A., Ather, S., Silva, C. F., Han, D., Heussel, C. P., Hickes, W., Kauczor, H.-U., Novotny, P., Peschl, H., Rook, M., Rubtsov, R., von Stackelberg, O., Tsakok, M. T., Arteta, C., Declerck, J., Kadir, T., Pickup, L., Gleeson, F., & Oudkerk, M. (2021). Lung cancer prediction by deep learning to identify benign lung nodules. *Lung Cancer*, 154, 1–4. https://doi.org/10.1016/j.lungcan.2021.01.027.

[41] Zadeh Shirazi, A., Fornaciari, E., McDonnell, M. D., Yaghoobi, M., Cevallos, Y., Tello-Oquendo, L., Inca, D., & Gomez, G. A. (2020). The application of deep convolutional neural networks to brain cancer images: A survey. *Journal of Personalized Medicine*, 10(4), 224. https://doi.org/10.3390/jpm10040224.

[42] Tandel, G. S., Balestrieri, A., Jujaray, T., Khanna, N. N., Saba, L., & Suri, J. S. (2020). Multiclass magnetic resonance imaging brain tumor classification using artificial intelligence paradigm. *Computers in Biology and Medicine*, 122, 103804. https://doi.org/10.1016/j.compbiomed.2020.103804.

[43] Abd-Ellah, M. K., Awad, A. I., Khalaf, A. A., & Hamed, H. F. (2018). Two-phase multi-model automatic brain tumour diagnosis system from magnetic resonance images using convolutional neural networks. *EURASIP Journal on Image and Video Processing*, 2018(1), 1–11. https://doi.org/10.1186/s13640-018-0332-4.

[44] Johnson, M., Albizri, A., & Simsek, S. (2020). Artificial Intelligence in healthcare operations to enhance treatment outcomes: A framework to predict lung cancer prognosis. *Annals of Operations Research*, 308(1–2), 275–305. https://doi.org/10.1007/s10479-020-03872-6.

[45] Sannasi Chakravarthy, S. R., & Rajaguru, H. (2019). Lung cancer detection using probabilistic neural network with modified crow-search algorithm. *Asian Pacific Journal of Cancer Prevention*, 20(7), 2159–2166. https://doi.org/10.31557/apjcp.2019.20.7.2159.

[46] Barker, J., Hoogi, A., Depeursinge, A., & Rubin, D. L. (2016). Automated classification of brain tumor type in whole-slide digital pathology images using local representative tiles. *Medical Image Analysis*, 30, 60–71. https://doi.org/10.1016/j.media.2015.12.002.

[47] Shyamsunder Shukla, R., & Aggarwal, Y. (2018). Nonlinear heart rate variability based artificial intelligence in lung cancer prediction. *Journal of Applied Biomedicine*, 16(2), 145–155. https://doi.org/10.1016/j.jab.2017.12.002.

[48] Selvanambi, R., Natarajan, J., Karuppiah, M., Islam, S. K. H., Hassan, M. M., & Fortino, G. (2018). Retracted article: Lung cancer prediction using higher-order recurrent neural network based on glowworm swarm optimization. *Neural Computing and Applications*, 32(9), 4373–4386. https://doi.org/10.1007/s00521-018-3824-3.

15 Internet of Vehicles (IoV)
Transforming Transportation through Connectivity, Big Data, and Security

*Ashima Arya, Prince Gupta, Jaya Sharma,
Sapna Juneja, and Swasti Singhal*

15.1 INTRODUCTION

The Internet of Vehicles (IoV) is a network of automobiles equipped with sensors, software, and other technologies that enable them to exchange data and communicate via the Internet [2]. By including additional communication channels including vehicle-to-vehicle (V2V), vehicle-to-infrastructure (V2I), vehicle-to-roadside (V2R), vehicle-to-person (V2P), vehicle-to-sensor (V2S), and vehicle-to-cloud (V2C), it expands the definition of vehicular ad hoc networks (VANETs) [1]. IoV is very important to the transportation sector. It might completely transform a number of transportation-related industries and provide a host of advantages.

Several significant issues to take into account include for IoV are:

Improved Safety: It enables real-time communication and information exchange between vehicles, infrastructure, and pedestrians, leading to enhanced safety on the roads. It enables the implementation of advanced driver assistance systems, collision avoidance mechanisms, and emergency response systems [3].

Efficient Traffic Management: It makes it easier for intelligent transportation systems to track and control traffic in real time. Reduced traffic congestion can be achieved by gathering and evaluating data from infrastructure and cars, which will improve route planning and traffic signal management.

Smart Mobility and Transportation Services: It enables the development of smart mobility solutions, such as ride-sharing, carpooling, and autonomous vehicles. It offers convenient transportation services, improved fleet management, and efficient logistics.

Environmental Benefits: Through IoV, vehicles can communicate and collaborate to optimize fuel efficiency, reduce emissions, and promote eco-friendly driving behaviors. It contributes to sustainable transportation practices and environmental conservation.

DOI: 10.1201/9781032673479-19

Advanced Vehicle Diagnostics and Maintenance: It allows vehicles to collect and transmit diagnostic data in real time. This enables proactive maintenance, early detection of faults, and remote software updates, leading to improved vehicle performance and reliability.

Enhanced User Experience: It provides a personalized and connected experience to vehicle occupants. It offers seamless integration with personal devices, entertainment systems, and smart home technologies, making travel more enjoyable and productive.

The development and implementation of IoV require addressing various challenges, including security, privacy, data management, standardization, and infrastructure requirements. However, with its potential to transform transportation, IoV is an exciting field that continues to evolve and innovate [4–7,23].

15.2 TECHNOLOGIES USED IN IOV

In order to establish a network of intelligent automobiles, Internet of automobiles (IoV) integrates mobile Internet with Internet of Things (IoT) [3] (Figure 15.1).

Mobile Internet: The mobile Internet refers to the connectivity and communication capabilities provided by mobile networks. It enables vehicles to connect to the Internet and communicate with other vehicles, infrastructure, and various online services. By utilizing mobile Internet connectivity, vehicles in the IoV can access real-time information, exchange data, and interact with the broader network.

Internet of Things (IoT): It is a network of connected physical objects, sensors, and devices that can collect and share data because they have Internet connectivity built-in. In the context of IoV, vehicles become an integral part of the IoT ecosystem. They have a range of sensors, software, and communication tools that let them produce and exchange data. Global Positioning System (GPS), cameras, radars, and internal car systems are a few examples of these sensors.

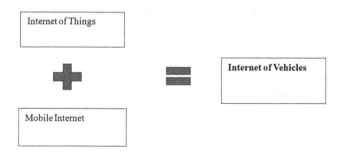

FIGURE 15.1 Mobile Internet and the Internet of Things are combined in the Internet of Vehicles (IoV).

15.3 WORKING OF NETWORK OF SMART VEHICLES MODEL (NSVM)

IoV establishes a network of intelligent vehicles that can connect with one another and with their surroundings by fusing the mobile Internet with IoT [10]. Here is how it works:

Vehicle Connectivity: IoV makes it possible for automobiles to communicate wirelessly to other vehicles, infrastructure (such traffic lights or road signs), and external services. Real-time information sharing and data exchange are made possible by this link.

Data Collection and Analysis: Smart vehicles in the IoV collect a vast amount of data from their onboard sensors and systems. This data includes information about vehicle performance, environmental conditions, traffic patterns, and more. The gathered data is processed and analyzed to yield insightful findings that facilitate improved judgment and intelligent functionalities.

Real-Time Communication: IoV enables in vehicle and other network entity real-time communication. Sharing of safety-related information, traffic updates, and cooperative driving are all made possible via communication between cars (V2V) and between vehicles and infrastructure (V2I). To improve traffic flow and improve road safety, vehicles can communicate information about their speed, location, and intended moves.

Intelligent Services and Applications: The IoV enables the creation of intelligent services and apps that enhance driving functionality. These can include real-time navigation and route optimization, adaptive cruise control, parking assistance, remote diagnostics, and vehicle-to-cloud services.

Integration with IoT Ecosystem: IoV integrates with the broader IoT ecosystem, allowing vehicles to interact with other IoT devices and services. For example, a smart vehicle can communicate with smart home systems, personal devices, and online platforms to exchange data, control home appliances, access personal preferences, and enable seamless experiences.

By combining the mobile Internet and IoT, IoV transforms vehicles into intelligent nodes within a larger sensing network. This connectivity and intelligence bring numerous benefits, including enhanced safety, improved traffic management, smart mobility services, and personalized user experiences.

Security, privacy, interoperability, and data management are still issues with IoV. The potential for IoV to revolutionize transportation and provide smarter, more effective vehicles and road networks is encouraging, though, as technology continues to advance (Figure 15.2).

15.4 IOV TECHNOLOGY AND ARCHITECTURE

The growth of the Internet of automobiles (IoV) has been made possible in large part by the connectivity between automobiles and the IoT [8]. The main features of this connectivity and its importance are discussed below:

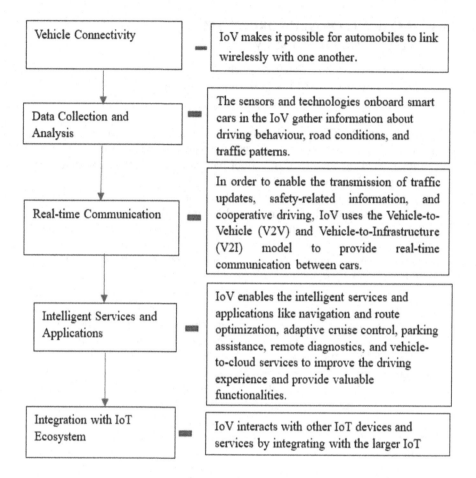

FIGURE 15.2 Working of NSVM.

Vehicle-to-Vehicle (V2V) Communication: The IoT connectivity allows vehicles to establish direct communication with other vehicles in their vicinity. Vehicle speed, position, acceleration, and other real-time information can be exchanged via V2V communication. This information sharing enhances road safety by enabling collision avoidance, cooperative driving, and traffic coordination.

Vehicle-to-Infrastructure (V2I) Communication: IoV relies on the connectivity between vehicles and infrastructure elements, as stop signs, traffic signals, and roadside sensors. Vehicles can interact with the surrounding infrastructure thanks to V2I communication, allowing them to receive traffic updates, signal prioritization, and information related to road conditions. This connectivity enhances traffic management, optimizes traffic flow, and enables intelligent transportation systems.

Vehicle-to-Cloud (V2C) Communication: IoV leverages cloud computing capabilities to enable advanced services and applications. Vehicles can connect to the cloud to access a range of services, including real-time

navigation, infotainment, remote diagnostics, software updates, and personalized settings. V2C communication allows vehicles to exchange data with
cloud-based platforms and services, enabling seamless connectivity and
enhancing the driving experience.

Vehicle-to-Personal (V2P) Communication: The interoperability of automobiles
and personal electronics like smartphones and wearables, is another important aspect of IoV. V2P communication enables personalized services, such as
remote vehicle monitoring and control, personalized in-car settings, and integration with personal calendars, contacts, and preferences. This connectivity
enhances convenience, comfort, and customization for vehicle occupants.

Vehicle-to-Sensor (V2S) Communication: IoV relies on the integration
of various sensors within vehicles. V2S communication enables vehicles
to connect and interact with external sensors, such as roadside sensors,
weather stations, and surveillance systems. This connectivity allows vehicles to gather additional data about the environment, road conditions, and
potential hazards, enhancing safety and situational awareness (Figure 15.3).

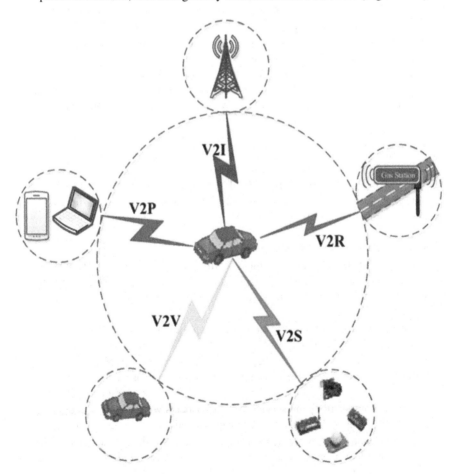

FIGURE 15.3   Vehicular communications of IoV.

The connectivity between vehicles and IoT in IoV brings several significant benefits:

a. **Enhanced Safety**: Real-time communication between vehicles enables the exchange of safety-related information, supporting collision avoidance and cooperative driving. This connectivity can prevent accidents, reduce congestion, and improve overall road safety.

b. **Improved Efficiency**: IoV connectivity enables optimized traffic management and coordination between vehicles and infrastructure. This improves traffic flow, reduces congestion, and enhances transportation efficiency.

c. **Intelligent Services**: The connectivity enables the development of intelligent services and applications, such as real-time navigation, predictive maintenance, remote diagnostics, and personalized in-car experiences. These services enhance convenience, comfort, and efficiency for vehicle occupants.

d. **Data-Driven Insights**: The connectivity between vehicles and IoT facilitates the collection and analysis of large amounts of data. This data can be utilized to derive valuable insights about traffic patterns, driver behavior, road conditions, and more. These insights can inform decision-making, urban planning, and the development of future mobility solutions.

e. **Seamless Integration**: IoV connectivity allows vehicles to seamlessly integrate with other IoT devices and services. This integration enables a holistic ecosystem where vehicles can interact with smart homes, smart cities, and other IoT-enabled environments, providing a seamless and connected experience for users.

15.4.1 ARCHITECTURE

The general architecture of the Internet of Vehicles (IoV) encompasses various components, including layered architecture, protocol stack, and network model. Here is an overview of each aspect:

15.4.1.1 Layered Architecture

IoV follows a layered architecture model, which organizes the system into distinct layers, each responsible for specific functionalities [9,12]. The layered architecture typically consists of the following layers:

a. **Application Layer**: This layer includes applications and services that provide specific functionalities to users and vehicles, such as navigation systems, infotainment, remote diagnostics, and vehicle management.

b. **Service Layer**: The service layer facilitates communication and coordination between different applications and services. It manages the service discovery, service composition, and service delivery aspects within the IoV ecosystem.

c. **Network Layer**: The network layer handles the routing and forwarding of data packets between vehicles, infrastructure, and the cloud. It establishes communication paths, manages network resources, and ensures efficient data transfer.

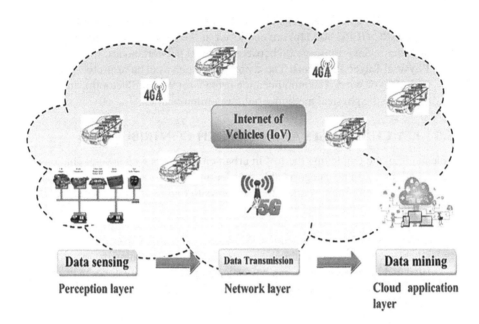

FIGURE 15.4 Layered architecture of IoV.

d. **Data Link Layer**: The data link layer handles the transmission of data between adjacent network nodes, typically within the same physical proximity. It ensures reliable and error-free data transfer through protocols such as Wi-Fi, Bluetooth, or Dedicated Short-Range Communications (DSRC).

e. **Physical Layer**: The physical layer deals with the actual transmission of data over physical mediums, such as wireless communication technologies (e.g., cellular networks, Wi-Fi, DSRC) or wired connections (e.g., Ethernet) (Figure 15.4).

15.4.2 Protocol Stack

The protocol stack in IoV refers to the set of communication protocols used at different layers of the architecture. Each layer in the architecture typically has its own set of protocols. Here are some common protocols used in the IoV protocol stack:

a. **Application Layer Protocols**: These protocols enable communication between applications and services, such as HTTP, MQTT, and CoAP for data exchange and RESTful APIs for service integration.

b. **Transport Layer Protocols**: Protocols like TCP and UDP provide reliable and connection-oriented or connectionless transport of data between endpoints.

c. **Network Layer Protocols**: Internet Protocol is commonly used in the network layer to route data packets between different network nodes. Other protocols like ICMP (Internet Control Message Protocol) may be used for network management and diagnostics.

d. **Data Link Layer Protocols**: Protocols like Ethernet, Wi-Fi (IEEE 802.11), and DSRC (IEEE 802.11p) are employed in the data link layer for efficient and secure data transmission between adjacent network nodes.

e. **Physical Layer Protocols**: These protocols, such as cellular protocols (4G/ LTE, 5G) or wireless communication protocols (Wi-Fi, Bluetooth), are particular to the physical medium used for communication.

15.5 KEY CHALLENGES AND RESEARCH CONTRIBUTIONS

Implementing and employing the IoV in urban cities comes with several challenges. These challenges can be categorized into different areas.

Here are some common challenges also shows in Figure 15.5 associated with IoV implementation in urban cities are [10]:

Connectivity and Communication: Vehicular Network Coverage: Ensuring reliable network coverage throughout the urban area, including densely populated and high-traffic areas, can be challenging. Connectivity issues may arise due to signal interference, network congestion, or gaps in network infrastructure.

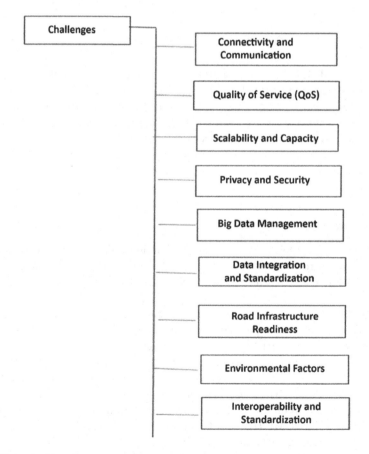

FIGURE 15.5 Key challenges.

Scalability and Capacity: As the number of connected vehicles increases, the network must scale to accommodate the growing traffic and communication demands. Ensuring sufficient network capacity and scalability is crucial for the seamless operation of IoV in urban environments.

Quality of Service (QoS): Providing consistent and reliable communication with acceptable levels of latency, bandwidth, and data transmission rates is essential for delivering high-quality services within IoV. QoS management becomes more challenging in congested urban areas with a high density of vehicles.

Big Data Management: IoV generates a vast amount of data from vehicles, infrastructure, and sensors. Efficient data collection, storage, processing, and analysis present significant challenges. Managing and extracting actionable insights from Big Data can be complex and resource-intensive.

Privacy and Security: Shielding delicate data and confirming the security of IoV systems and communication networks is crucial. Safeguarding against unauthorized access, data breaches, cyber-attacks, and ensuring data privacy are ongoing challenges for IoV implementation in urban cities.

Data Integration and Standardization: Integrating data from diverse sources, such as vehicles, infrastructure, and external sensors, poses challenges in terms of data formats, protocols, and interoperability. Developing standardized data formats and communication protocols is essential for seamless data integration and exchange.

Road Infrastructure Readiness: The existing road infrastructure in urban cities may not be fully equipped to support IoV implementation. Deploying necessary infrastructure elements like sensors, traffic management systems, and communication infrastructure requires significant investment and coordination with city authorities.

Environmental Factors: Urban environments present unique challenges due to factors such as high-traffic density, urban canyons (tall buildings obstructing wireless signals), and environmental conditions (e.g., weather and pollution). These factors can impact the reliability and performance of IoV systems and communication.

Policy and Regulations: Establishing a regulatory framework that governs the operation, data privacy, liability, and safety aspects of IoV in urban cities is crucial. Addressing legal and regulatory challenges associated with data ownership, sharing, and usage is necessary for widespread adoption and implementation.

Interoperability and Standardization: Ensuring interoperability and standardization across different IoV systems, applications, and communication protocols is essential for seamless operation and collaboration between various stakeholders. Establishing common standards and protocols can facilitate interoperability and enhance the effectiveness of IoV in urban cities.

These challenges require collaborative efforts among stakeholders, including governments, city planners, technology providers, and automotive industry players, to address them effectively and enable the successful implementation and deployment of IoV in urban cities.

The field of IoV has witnessed significant research contributions and advancements. Here are some notable research contributions and advancements in the field [16–21]:

Development of V2V and V2X Communication: Research has focused on developing efficient and reliable V2V and V2X communication protocols. These advancements enable vehicles to exchange information with other vehicles, infrastructure, pedestrians, and the cloud, improving safety, traffic management, and overall efficiency [1].

Integration with 5G and Beyond: The amalgamation of IoV with evolving wireless communication technologies like 5G and beyond has been an area of active research. Utilizing 5G networks' high data speeds, low latency, and widespread device connectivity improves the functionality and efficiency of IoV applications [2].

The field of IoV has witnessed significant research contributions and advancements. Here are some notable research contributions and advancements in the field [16–21]:

Development of V2V and V2X Communication: Research has focused on developing efficient and reliable V2V and V2X communication protocols. These advancements enable vehicles to exchange information with other vehicles, infrastructure, pedestrians, and the cloud, improving safety, traffic management, and overall efficiency [1].

Integration with 5G and Beyond: The integration of IoV with emerging wireless communication technologies like 5G and beyond has been an area of active research. Utilizing 5G networks' high data speeds, low latency, and widespread device connectivity improves the functionality and efficiency of IoV applications [2].

Big Data Analytics: Researchers have explored techniques for efficiently managing and analyzing the massive volume of data generated by IoV systems. In order to extract useful insights from vehicle data and enable a variety of applications including traffic prediction, anomaly detection, and intelligent routing, advanced analytics techniques like machine learning and data mining are used [3].

Edge Computing and Fog Computing: In order to overcome the difficulties presented by real-time processing and applications that require low latency, research has focused on utilizing edge computing and fog computing architectures. By bringing computing and storage closer to the network edge, these systems decrease the need for centralized cloud infrastructure and enable faster data processing [4].

Secure Communication and Authentication: Research has been dedicated to developing secure communication protocols and authentication mechanisms to protect the integrity and privacy of IoV systems. Encryption techniques, digital signatures, and secure key management are employed to ensure secure and trusted communication among vehicles and infrastructure [4].

Privacy-Preserving Techniques: Various privacy-preserving techniques have been explored to address concerns related to personal data privacy in IoV. Approaches like anonymous authentication, pseudonymization, and differential privacy techniques help protect user identities and sensitive information while still enabling effective data sharing and analysis [1].

Autonomous Vehicle Technologies: Research advancements in IoV have contributed to the development of autonomous vehicle technologies. These include sensor fusion, perception algorithms, path planning, and decision-making techniques, enabling vehicles to navigate and interact with their environment autonomously [6].

Cooperative and Intelligent Transportation Systems: Research has focused on cooperative and intelligent transportation systems that leverage IoV technologies. These technologies allow for communication between auto-mobiles and the infrastructure, which improves traffic flow, lessens conges-tion, and increases safety [5].

These research contributions and advancements have significantly advanced the field of IoV, paving the way for innovative applications, improved road safety, enhanced traffic management, and a more efficient transportation ecosystem. Continued research in IoV will further drive its evolution and bring about transformative changes in the transportation industry (Figure 15.6).

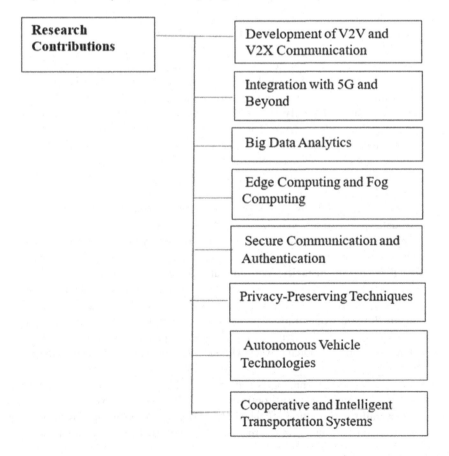

FIGURE 15.6 Contributions to research and developments in the field.

15.6 BIG DAT IN IOV

Big Data plays a crucial role in the field of IoV by enabling the collection, storage, processing, and analysis of large volumes of data generated by vehicles and associated systems.

Here are some key aspects of Big Data in IoV:

Data Acquisition: Massive volumes of data are produced by IoV systems from a variety of sources, including vehicle sensors, GPS devices, cameras, radars, and external sensors. This data includes real-time vehicle information, environmental data, traffic data, and driver behavior data. Efficient data acquisition techniques are required to gather and aggregate this data from multiple vehicles and sensors.

Data Storage: Big Data technologies and architectures are employed to store and manage the vast amount of data generated by IoV systems. The scalability and reliability demands of IoV data storage are handled by distributed storage solutions like Hadoop Distributed File System (HDFS) and cloud storage. Additionally, technologies like data replication and fault tolerance mechanisms ensure data availability.

Data Processing: Big Data processing techniques are applied to extract valuable insights from the collected data. Real-time data processing frameworks, such as Apache Kafka and Apache Storm, enable the analysis of streaming data for immediate decision-making. Batch processing frameworks like Apache Spark and Hadoop MapReduce are used for large-scale data analysis, data mining, and predictive analytics.

Data Analysis and Analytics: Advanced analytics methods are applied to perform in-depth analysis of IoV data. Machine learning algorithms and statistical techniques are used to identify patterns, trends, and anomalies in the data. This enables applications such as traffic prediction, congestion detection, predictive maintenance, and intelligent routing based on historical and real-time data.

Data Integration and Fusion: IoV systems often require the integration and fusion of data from multiple vehicles, sensors, and infrastructure. Data integration approaches guarantee the smooth exchange of information across various IoV ecosystem components, facilitating thorough analysis and decision-making.

Data Privacy and Security: With the large amount of sensitive and personal data being collected in IoV, ensuring data privacy and security is of paramount importance. To prevent unwanted access or misuse of data, access control methods, secure data transfer protocols, and encryption techniques are used. Anonymization and differential privacy are two privacy-preserving strategies that safeguard individual privacy while facilitating data sharing and analysis.

By leveraging Big Data technologies and analytics in IoV, stakeholders can gain valuable insights into traffic patterns, optimize route planning, improve road safety, and enhance overall transportation efficiency. Big Data in IoV opens up possibilities for innovative applications and services that can revolutionize the transportation industry.

15.6.1 TECHNOLOGIES USED IN BIG DATA FOR IoV

There are several technologies and approaches that play a crucial role in handling Big Data in the context of IoV. These technologies enable acquisition, transformation, storage, processing, analysis, and decision-making processes associated with Big Data in IoV.

Here are some notable technologies for Big Data in IoV:

1. **Data Acquisition:**
   - **Sensor Networks**: Utilize various sensors installed in vehicles to collect data on location, speed, acceleration, temperature, fuel consumption, and other relevant parameters.
   - **Global Positioning System (GPS)**: Provides precise positioning information for vehicles, enabling location-based data collection.
2. **Data Transformation and Normalization:**
   - **Data Brokers**: Facilitate the integration and aggregation of heterogeneous data from different sources in IoV, ensuring compatibility and consistency.
   - **Data Preprocessing**: Involves data cleaning, filtering, and transformation to enhance data quality and prepare it for further analysis.
3. **Data Storage:**
   - **Distributed Storage Systems**: Enable scalable and fault-tolerant storage of large volumes of IoV data, such as distributed file systems (e.g., Hadoop) or NoSQL databases (e.g., Apache Cassandra).
   - **Cloud Storage**: Provides on-demand, elastic storage capabilities for IoV data, allowing seamless scalability and accessibility.
4. **Data Processing:**
   - **Real-time Processing**: Involves analyzing and processing streaming data in real time, enabling immediate responses and timely decision-making.
   - **Batch Processing**: Utilizes parallel processing frameworks like Apache Spark or Hadoop MapReduce to handle large-scale batch data processing tasks.
5. **Data Analysis:**
   - **Data Mining**: Extracts patterns, trends, and insights from large datasets to discover useful information and knowledge.
   - **Machine Learning**: Utilizes algorithms and statistical models to enable predictive analytics, anomaly detection, and pattern recognition in IoV data.
6. **Decision-Making:**
   - **Decision Support Systems**: Utilize advanced analytics and visualization techniques to assist in making informed decisions based on analyzed IoV data.
   - **Intelligent Transportation Systems**: Employ data-driven insights to optimize traffic management, route planning, resource allocation, and overall transportation efficiency.

7. **Data Security and Privacy**:
   - **Encryption and Access Control**: Protects sensitive IoV data through encryption techniques and controls access to ensure data security and privacy.
   - **Privacy-Preserving Data Analytics**: Implements techniques like differential privacy or secure multi-party computation to perform data analysis while preserving individual privacy.

These technologies collectively enable the effective handling and utilization of Big Data in IoV, allowing for improved vehicle performance, enhanced traffic management, optimized resource allocation, and the development of innovative applications and services in the transportation industry.

15.7 SECURITY AND PRIVACY IN IOV

IoV is revolutionizing the transportation industry by creating a network of connected and smart vehicles. Nonetheless, serious security and privacy concerns are brought up by the IoV's broad adoption. This study examines the security and privacy concerns related to the IoV and looks into suggested fixes and protocols to lessen these difficulties [11–15].

Within the Internet of automobiles (IoV) ecosystem, automobiles are linked to a multitude of entities, such as other vehicles, infrastructure systems, and the IoT. Due to the risks brought forth by this connectivity, vehicle systems and data availability, confidentiality, and integrity may be jeopardized by hostile actors. Additionally, privacy concerns arise due to the collection and sharing of sensitive information, including location data, driving patterns, and personal identifiable information.

To address these security and privacy challenges, researchers and industry experts have proposed several solutions. These include secure communication protocols, authentication mechanisms, intrusion detection systems, and encryption techniques. This study presents an overview of the security and privacy issues in the IoV, classifies the difficulties encountered, and investigates the suggested protocols and remedies. It examines the importance of end-to-end security, secure vehicle-to-vehicle and vehicle-to-infrastructure communication, and privacy-preserving techniques. The paper also discusses the need for standardized security frameworks and regulations to ensure a secure and privacy-enhancing IoV environment.

15.7.1 THE SECURITY AND PRIVACY ISSUES IN IoV

IoV brings numerous benefits to the transportation industry, but it also introduces security and privacy challenges that need to be addressed for the widespread adoption and success of IoV systems. Few most important security and privacy concerns in IoV will be covered in this section.

Unauthorized Access and Attacks: IoV systems are vulnerable to various types of attacks, including unauthorized access, hacking, and cyber-attacks. Malicious actors can exploit vulnerabilities in vehicle communication protocols, software, and hardware to gain unauthorized control over vehicles or disrupt their operations. Such attacks can lead to accidents, theft of sensitive information, or even physical harm to passengers.

Data Security and Privacy: IoV generates a massive amount of data, including location information, vehicle telemetry, and personal identifiable information. Ensuring the security and privacy of this data is crucial. Unauthorized access to this data can result in privacy breaches, identity theft, or misuse of personal information. Additionally, data integrity and confidentiality need to be protected to prevent tampering or unauthorized modifications that can affect the safety and reliability of vehicles.

Vehicular Network Security: The communication networks used in IoV, such as vehicle-to-vehicle (V2V) and vehicle-to-infrastructure (V2I) networks, are prone to security threats. These networks can be targets of eavesdropping, man-in-the-middle attacks, jamming, or message spoofing. To maintain a secure IoV environment, it is crucial to guarantee the sincerity, reliability, and secrecy of communications inside the vehicular network.

Privacy Concerns and Tracking: Massive volumes of data, such as location, driving habits, and personal preferences, are gathered and processed by IoV systems. This raises privacy concerns as individuals may not want their movements and behavior to be continuously monitored and tracked. It is crucial to implement privacy-preserving mechanisms to protect the anonymity of individuals while still enabling efficient and secure data sharing for legitimate purposes.

System Integration and Interoperability: IoV involves the integration of various systems, including vehicles, infrastructure, and IoT devices. Ensuring secure and seamless interoperability among these systems is a challenge. The diverse technologies, protocols, and standards used in different IoV components need to be harmonized and validated to mitigate security risks and maintain the integrity of the overall system.

To address these security and privacy challenges, researchers and industry practitioners are working on various solutions. These include the development of secure communication protocols, encryption techniques, intrusion detection systems, access control mechanisms, and robust authentication and authorization frameworks. Additionally, emerging technologies like blockchain are being explored to enhance the security and privacy of IoV systems through decentralized and tamper-resistant data management.

15.7.2 PROPOSED SOLUTIONS AND PROTOCOLS FOR SECURITY AND PRIVACY IN IoT

Many suggested protocols and solutions are being developed to solve the security and privacy concerns in the IoT, especially as they relate to the use of IoV. Here are some of the notable solutions and protocols:

Secure Communication Protocols: Numerous protocols have been developed to provide safe communication between IoT devices and systems. Two such protocols include the Datagram Transport Layer Security (DTLS) protocol, which is intended for limited IoT devices, and the Transport Layer Security (TLS) protocol, which offers end-to-end encryption and authentication. These protocols establish secure connections, protect data integrity, and verify the authenticity of communication partners.

Lightweight Cryptography: Lightweight cryptographic techniques have been suggested to enable effective and safe data encryption because IoT devices frequently have limited processing power and memory. These techniques, such Elliptic Curve Cryptography (ECC) and Lightweight Cryptography (LWC), strike a compromise between resource efficiency and security, making them appropriate for IoT devices with limited resources.

Identity and Access Management: To mitigate unauthorized access and ensure proper authentication and authorization, identity and access management mechanisms are crucial. Protocols like OAuth and OpenID Connect are used to enable secure authentication and authorization between IoT devices, applications, and services. These protocols allow for controlled access to IoT resources and protect against unauthorized usage.

Intrusion Detection Systems (IDSs): IDSs are designed to detect and respond to potential security breaches or malicious activities in IoT networks. These systems monitor network traffic, analyze data patterns, and raise alerts or take preventive actions when anomalies or suspicious activities are detected. Machine learning and anomaly detection techniques are commonly employed to enhance the accuracy and effectiveness of IDSs.

Blockchain Technology: Blockchain has gained attention as a potential key to boost security and privacy in IoT. It offers decentralized and tamper-resistant data storage, consensus mechanisms, and smart contracts. In IoV, blockchain can be utilized to secure vehicle data, establish trust between participants, enable secure and transparent data sharing, and prevent tampering or unauthorized modifications.

Privacy-Preserving Techniques: A number of privacy-preserving methods have been put out to safeguard users' and IoT data's privacy. One such method is differential privacy, which ensures individual privacy by adding noise or unpredictability to the data collected while allowing for meaningful analysis. Homomorphic encryption enables performing computations on encrypted data without decrypting it, preserving data privacy.

Secure Firmware and Software Updates: Ensuring the security of IoT devices throughout their lifecycle is crucial. Secure firmware and software update mechanisms are employed to patch vulnerabilities, fix bugs, and deploy security patches to IoT devices in a secure manner. Over-the-air (OTA) update protocols and secure boot mechanisms are used to prevent unauthorized or malicious updates.

15.7.3 COMPARISON OF SOME KEY SOLUTIONS AND PROTOCOLS

Secure Communication Protocols:

- **TLS (Transport Layer Security)**: Provides end-to-end encryption and authentication, ensuring secure communication between IoV devices and systems.
- **DTLS (Datagram Transport Layer Security)**: Designed for resource-constrained IoV devices, it establishes secure connections while considering their limited capabilities.

Lightweight Cryptography:

- **ECC (Elliptic Curve Cryptography)**: Offers efficient encryption and authentication mechanisms suitable for resource-constrained IoV devices.
- **LWC (Lightweight Cryptography)**: Provides cryptographic algorithms optimized for low-power devices, balancing security and resource efficiency.

Identity and Access Management:

- **OAuth (Open Authorization)**: Enables secure authentication and authorization between IoV devices, applications, and services, ensuring controlled access to resources.
- **OpenID Connect**: Facilitates single sign-on and identity federation, allowing users to authenticate across different IoV systems securely.

Intrusion Detection Systems (IDS):

- **IoT IDS**: Specifically designed for detecting and responding to security threats in IoV networks, employing techniques like anomaly detection and machine learning to identify malicious activities.

Blockchain Technology:

- **Immutable Ledger**: Offers a tamper-resistant and transparent record of transactions, enhancing data security and integrity in IoV systems.
- **Smart Contracts**: Enables secure and automated execution of predefined rules and agreements, providing trust and reducing reliance on centralized authorities.

Privacy-Preserving Techniques:

- **Differential Privacy**: Protects individual privacy by adding noise or randomness to data, allowing for meaningful analysis without revealing sensitive information.
- **Homomorphic Encryption**: Ensures data privacy in IoV applications by allowing computations to be performed on encrypted data without first decrypting it.

Secure Firmware and Software Updates:

- **OTA (Over-the-Air) Updates**: Ensures secure and authenticated firmware and software updates for IoV devices, preventing unauthorized modifications.
- **Secure Boot**: Verifies the integrity and authenticity of device firmware during the boot-up process, preventing unauthorized or malicious software execution.

While these solutions and protocols address security and privacy in IoV, it is essential to evaluate their applicability, efficiency, and scalability based on specific use cases and deployment scenarios. Ongoing research and collaboration among academia, industry, and standardization bodies aim to further enhance the security and privacy of IoV systems and mitigate emerging threats.

15.8 FUTURE DIRECTIONS AND EMERGING TECHNOLOGIES IN IOV

The field of the IoV continues to evolve rapidly, and several future directions and emerging technologies hold promise for further advancements in this domain. These developments aim to enhance the capabilities, efficiency, and safety of IoV systems [19–23].

5G and Beyond: The deployment of 5G networks and the upcoming evolution to 6G will significantly impact IoV by providing ultra-low latency, high bandwidth, and massive device connectivity. Real-time V2V and V2I communication will be made possible as a result, opening the door to autonomous driving, better traffic management, and increased safety.

Edge Computing: Edge computing puts processing power closer to the data source in response to the growing amount of data created by IoV devices. This lessens reliance on cloud infrastructure, improves real-time decision-making, and lowers latency. Edge computing enables faster response times, efficient data processing, and improved privacy by keeping sensitive data localized.

Artificial Intelligence (AI) and Machine Learning (ML): AI and ML technologies will play a crucial role in IoV by enabling intelligent decision-making, predictive analytics, and autonomous vehicle operations. Large amounts of IoV data can be analyzed by AI algorithms to find patterns, forecast traffic situations, plan routes, and improve system performance.

Blockchain Technology: Blockchain allows for decentralized and transparent data management, which can improve the IoV systems' security, privacy, and dependability. It can facilitate secure and tamper-proof vehicle-to-vehicle communication, data sharing, and transactional activities, such as secure payments and data exchange.

Vehicular Edge Computing: Vehicular edge computing leverages the computing and storage capabilities of vehicles to process data locally. It enables efficient data offloading, collaborative processing among neighboring vehicles, and localized decision-making. Vehicular edge computing reduces network congestion, improves response times, and enhances privacy by minimizing data transmission to centralized servers.

Cooperative Intelligent Transportation Systems (C-ITS): C-ITS involves the integration of vehicles, infrastructure, and various stakeholders to improve transportation efficiency and safety. C-ITS technologies enable cooperative sensing, communication, and decision-making among vehicles and infrastructure, leading to optimized traffic flow, collision avoidance, and emergency response.

Vehicle-to-Everything (V2X) Communication: V2X communication encompasses various forms of communication, including V2V, V2I, V2P (vehicle-to-pedestrian), and V2C (vehicle-to-cloud). The development and deployment of robust and standardized V2X communication protocols will enable seamless and secure connectivity among vehicles, infrastructure, pedestrians, and cloud-based services.

Cybersecurity and Privacy Enhancements: As IoV systems become more complex and interconnected, ensuring robust cybersecurity measures and privacy protection becomes critical. Future advancements will focus on developing advanced encryption techniques, intrusion detection systems, secure authentication mechanisms, and privacy-preserving technologies to safeguard IoV systems from cyber threats and protect user privacy.

These future directions and emerging technologies in IoV hold immense potential to transform transportation systems, enhance road safety, improve traffic management, and revolutionize the way vehicles interact with each other and the surrounding infrastructure. Continued research and development in these areas will shape the future of IoV and pave the way for a smarter and more connected transportation ecosystem.

15.9 CONCLUSION AND FUTURE SCOPE

IoV emerges as a transformative technology with significant implications for the transportation industry. The fusion of the mobile Internet and the IoT promotes driving comfort, efficiency, and safety by enabling the formation of a network of smart cars. However, challenges remain in terms of implementation, security, and privacy. With ongoing research and advancements, IoV holds immense potential for shaping the future of transportation. As IoV continues to evolve and integrate with emerging technologies, it is crucial to address the security and privacy concerns to build trust among users and stakeholders. This research highlights the advancements in security and privacy within the IoV domain and serves as a foundation for future studies in developing robust and resilient IoV systems.

It is important to note that security and privacy in IoT are ongoing research areas, and new solutions and protocols continue to emerge. Standardization bodies, industry alliances, and research organizations are actively working toward establishing best practices and guidelines to enhance the security and privacy of IoT systems, including those deployed in IoV.

REFERENCES

1. S. El Madani, S. Motahhir and A. El Ghzizal, "Internet of Vehicles: Concept, Process, Security Aspects and Solutions," *Multimedia Tools and Applications*, vol. 81, pp. 16563–16587, 2022, doi:10.1007/s11042-022-12386-1.
2. L.-M. Ang, K. P. Seng, G. K. Ijemaru and A. M. Zungeru, "Deployment of IoV for Smart Cities: Applications, Architecture, and Challenges," *IEEE Access*, vol. 7, pp. 6473–6492, 2019, doi:10.1109/ACCESS.2018.2887076.

3. A. D. Hansen, P. Sørensen, F. Iov and F. Blaabjerg, "Centralised Power Control of Wind Farm with Doubly Fed Induction Generators," *Renewable Energy,* vol. 31, no. 7, pp. 935–951, 2006, ISSN 0960-1481, doi:10.1016/j.renene.2005.05.011.

4. N. Aung, T. Kechadi, T. Zhu, S. Zerdoumi, T. Guerbouz and S. Dhelim, "Blockchain Application on the Internet of Vehicles (IoV)," *2022 IEEE 7th International Conference on Intelligent Transportation Engineering (ICITE),* Beijing, China, 2022, pp. 586–591, doi:10.1109/ICITE56321.2022.10101404.

5. W. Zhang, N. Aung, S. Dhelim and Y. Ai, "DIFTOS: A Distributed Infrastructure-Free Traffic Optimization System Based on Vehicular Ad Hoc Networks for Urban Environments," *Sensors (Switzerland),* vol. 18, no. 8, 1–18, 2018.

6. U. Javaid, M. N. Aman and B. Sikdar, "A Scalable Protocol for Driving Trust Management in Internet of Vehicles with Blockchain," *IEEE Internet Things Journal,* 7, pp. 1–1, 2020.

7. W. Wang, H. Ning, F. Shi, S. Dhelim, W. Zhang and L. Chen, "A Survey of Hybrid Human-Artificial Intelligence for Social Computing," *IEEE Transactions on Human-Machine Systems,* 52, 468–480, 2021.

8. J. Contreras-Castillo, S. Zeadally and J. A. Guerrero-Ibañez, "Internet of Vehicles: Architecture, Protocols, and Security," *IEEE Internet of Things Journal,* vol. 5, no. 5, pp. 3701–3709, 2018, doi:10.1109/JIOT.2017.2690902.

9. J. Guerrero-Ibanez, C. Flores-Cortes and S. Zeadally, "Vehicular Ad-Hoc Networks (VANETs): Architecture Protocols and Applications," In Naveen Chilamkurti, Sherali Zeadally, Hakima Chaouchi (Eds): *Next Generation Wireless Technologies: 4G and Beyond.* London, UK: Springer, 49–70 2013.

10. J. A. Guerrero-Ibanez, S. Zeadally and J. Contreras-Castillo, "Integration Challenges of Intelligent Transportation Systems with Connected Vehicle Cloud Computing and Internet of Things Technologies," *IEEE Wireless Communications,* vol. 22, no. 6, pp. 122–128, 2015.

11. L. Ben Othmane, H. Weffers, M. M. Mohamad and M. Wolf, *A Survey of Security and Privacy in Connected Vehicles,* New York: Springer, pp. 217–247, 2015.

12. Z. Mahmood, "Connected Vehicles in the IoV: Concepts, Technologies and Architectures." In: Mahmood, Z. (ed) *Connected Vehicles in the Internet of Things.* Cham: Springer, 2020. doi:10.1007/978-3-030-36167-9_.

13. F. Yang, S. Wang, J. Li, Z. Liu and Q. Sun, "An Overview of Internet of Vehicles," *China Communications,* vol. 11, no. 10, pp. 1–15, 2014, doi:10.1109/CC.2014.6969789.

14. W. Duan, J. Gu, M. Wen, G. Zhang, Y. Ji and S. Mumtaz, "Emerging Technologies for 5G-IoV Networks: Applications, Trends and Opportunities," *IEEE Network,* vol. 34, no. 5, pp. 283–289, 2020, doi:10.1109/MNET.001.1900659.

15. S. Parkvall., Dahlman, E., Furuskar, A. and Frenne, M et al., "NR: The New 5G Radio Access Technology," *IEEE Communications Standards Magazine,* vol. 1, no. 4, pp. 24–30, 2017.

16. O. Kaiwartya., Abdullah, A.H., Cao, Y., Altameem, A., Prasad, M., et al., "Internet of Vehicles: Motivation, Layered Architecture, Network Model, Challenges, and Future Aspects," *IEEE Access,* vol. 4, pp. 5356–5373, 2016, doi:10.1109/ACCESS.2016.2603219.

17. M.-S. Chen, C.-P. Hwang, T.-Y. Ho, H.-F. Wang, C.-M. Shih, H.-Y. Chen and W. K. Liu, "Driving Behaviors Analysis Based on Feature Selection and Statistical Approach: A Preliminary Study," *The Journal of Supercomputing,* vol. 75, pp. 2007–2026, 2019, doi:10.1007/s11227-018-2618-9.

18. L. Petersen, F. Iov and G. C. Tarnowski, "A Model-Based Design Approach for Stability Assessment, Control Tuning and Verification in Off-Grid Hybrid Power Plants," *Energies,* vol. 13, p. 49, 2020, doi:10.3390/en13010049.

19. Cheng., Yuan, G., Zhou, M., Gao, S., et al., "Accessibility Analysis and Modeling for IoV in an Urban Scene," *IEEE Transactions on Vehicular Technology*, vol. 69, no. 4, pp. 4246–4256, 2020, doi:10.1109/TVT.2020.2970553.

20. X. Lin, J. Wu, S. Mumtaz, S. Garg, J. Li and M. Guizani, "Blockchain-Based On-Demand Computing Resource Trading in IoV-Assisted Smart City," *IEEE Transactions on Emerging Topics in Computing*, vol. 9, no. 3, pp. 1373–1385, 2021, doi:10.1109/TETC.2020.2971831.

21. F. Sakiz and S. Sen, "A Survey of Attacks and Detection Mechanisms on Intelligent Transportation Systems: VANETs and IoV," *Ad Hoc Networks*, vol. 61, pp. 33–50, 2017, ISSN 1570-8705, doi:10.1016/j.adhoc.2017.03.006.

22. A. Hammoud, H. Sami, A. Mourad, H. Otrok, R. Mizouni and J. Bentahar, "AI, Blockchain, and Vehicular Edge Computing for Smart and Secure IoV: Challenges and Directions," *IEEE Internet of Things Magazine*, vol. 3, no. 2, pp. 68–73, 2020, doi:10.1109/IOTM.0001.1900109.

23. X. Krasniqi and E. Hajrizi, "Use of IoT Technology to Drive the Automotive Industry from Connected to Full Autonomous Vehicles", *IFACPapersOnLine*, vol. 49, no. 29, pp. 269–274, 2016.

16 A Secure, Scalable, and Integrated Smart Platform for Teaching and Coding

Mohit Dua, Giri Sainath Reddy, Ramesh Vishnoi, Satyam Tomar, and Shelza Dua

16.1 INTRODUCTION

The COVID-19 pandemic has significantly accelerated the need for online teaching, as it forced schools and universities to shut down their physical classrooms and move their instruction online. However, even before the pandemic, there was a growing demand for online teaching due to its many benefits. Online teaching, augmented by Internet of Things (IoT) technology, allows for greater accessibility, flexibility, and personalization, as students can learn from anywhere with an Internet connection and on their own schedule. It also provides a platform for collaboration and communication between students and teachers, leveraging IoT devices and connectivity to facilitate seamless interactions. IoT-enabled online teaching offers access to a wealth of educational resources, making it easier for students to explore digital content and engage with real-world data. Additionally, online teaching, supported by IoT, prepares students for a rapidly changing job market, where digital skills are becoming increasingly important. Overall, the need for online teaching, coupled with the integration of IoT technology, is driven by a desire for greater accessibility, flexibility, and innovation in education, as well as readiness for the digital era's evolving demands.

The online teaching and coding platform, enhanced with IoT technology and equipped with coding functionality, examination system, group video conferencing, and notes manager features has several advantages over existing apps. Firstly, its integration of IoT in teaching is a significant advantage. With the IoT-enabled coding platform feature, students can practice coding exercises and projects, and teachers can provide feedback and support in real time, leveraging IoT devices and sensors for enhanced interactivity. This integration ensures that coding is an integral part of the learning experience. Secondly, the platform prioritizes security and scalability. With the use of the MERN stack and Redux as state management, the platform provides a scalable and secure framework for building complex web applications. Additionally, the platform incorporates security features such as encryption, authentication, and access controls, ensuring that user data is protected while harnessing IoT capabilities for a more advanced online teaching experience.

The online teaching and coding platform that has coding platform, examination system, group video conferencing, and notes manager features is made using the

DOI: 10.1201/9781032673479-20

MERN stack with Redux as state management. The MERN stack is chosen because it provides a scalable and robust framework for building complex web applications. Additionally, Redux is used as a state management tool because it provides a centralized way to manage the state of the application, making it easier to debug and maintain. The use of Redux also ensures that data is consistent across components, reducing the risk of errors and improving performance.

The contributions of this chapter are summarized as follows:

1. Our work is to first focus on Building a platform that includes both coding and teaching system.
2. Later, our focus is to increase scalability and security in this app by state management and tokens.
3. We proposed an integrated idea of Teaching and Coding with keeping all necessary features and functionalities in mind.

The remaining part of the chapter is laid out as follows. The related work is presented in Section 16.2. Section 16.3 describes the proposed system architecture in detail. Results and discussion are given in Section 16.4 of the chapter, whereas Section 16.5 concludes the chapter.

16.2 RELATED WORK

There are several online teaching and coding platforms that exist today, catering to a wide range of audiences, from beginners to advanced learners. Gabriela Carmen Opriou [1] discussed that the growing importance of computer-based tools in the learning process is attributed to the emergence and widespread use of the Internet and its applications. There are several online teaching platforms like moodle, google classroom, etc. Maria Luminita Gotham [2] referred Moodle, as Modular Object-Oriented Dynamic Learning Environment, and is a widely recognized open-source Learning Management System and course management system, or Virtual Learning Environment (VLE). In Chun-Che Huang's [3] study, he drew the conclusions from two aspects: technology acceptance behavior and learning achievement. In terms of technology acceptance behavior, the study confirmed the following:

- Positive performance expectations significantly impact behavioral intentions.
- Social influence has a positive and significant impact on behavioral intentions.
- Help condition has a positive and significant impact on behavioral intentions.
- Gender and performance expectations adjustment variables have a significant impact on the path from behavioral intentions to performance.

According to Shampa Iftakhar et al. [4], teachers have been receptive to new technologies to facilitate optimal learning in both traditional and virtual classrooms. Mulyani et al. [5] have suggested that higher educational institutions have taken necessary steps to ensure that learning can continue effectively. This has been done by providing alternative, flexible, and effective learning strategies, methods, and techniques. As per Zinovieva et al. [6], a techno-ethical audit was done on Google Meet. The findings revealed a system that has restricted meaningful interaction, viewed students as passive technology users with minimal control or agency, and subjected them to excessive surveillance and monitoring practices with severe consequences.

Zinovieva et al. [6] have made a comparison of various online platforms used for teaching programming based on different criteria and found the hackerrank.com platform that they had previously used for teaching students. In [7], there has been a call for a broader conversation about how these categories can shape a comprehensive framework for online learning research, which can prioritize research aims over time and synthesize effects. As per [8], a study has suggested that although research on online teacher roles and competencies has informed the development of teacher training programs, further investigation is still needed. He has proposed an alternative perspective that views online teachers as adult learners who constantly transform their understanding of online teaching structures.

In [8], optimism has been expressed toward the potential of online education, suggesting that we have been entering a unique and exciting era, and that the previous challenges may be less daunting than before. Reference [9] says there may be no limit to experiential learning within online learning as long as faculty learns how to design experiential learning exercises that are universal. The authors in [10] have said the faculty participants in this study indicated that they expected a well-structured course, with students who have been proficient in using technology and submitting their work on time. The success of online teaching has largely depended on the instructor's ability to be energetic and communicative with students. In [11], it has been said that we have been in an exciting and unique period in the realm of online teaching and learning. The author believes that the challenges and difficulties that once posed a threat to the success of online education have been dissipating, making way for a brighter future in the field.

In higher education, students have always been the main clients [12], which has not been a new concept [13,14]. In [15], it has been highlighted that student satisfaction is the sole indicator for measuring service quality in higher education. The quality of service in higher education has had a significant impact on student satisfaction [16]. As competition among higher education institutions has continued to grow regionally, nationally, and globally, the notion of students as the main client has gained acceptance [17]. This has meant that universities have needed to develop ways to satisfy their students at this level of education [18,19], and this has been achieved by improving the quality of service they offer [20,21]. In online learning, it has been crucial to keep students motivated and minimize their frustration with this new mode of education [22]. However, as García-Peñalvo et al. [23] have indicated, assessment has been a complex process that has needed to be performed throughout a fixed instruction/learning period, not just at specific moments during the course. The attribute that has required attention in order to improve students' satisfaction with online teaching has been the usefulness of the system. When the virtual instruction system has functioned well, students have saved time in the learning process, it has boosted their sense of learning self-sufficiency, and has improved results [24–27].

Yingjie Shen, in his paper [28], discussed the application of the IoT in the modern education system using Android Voice assistance. Jasim, in [29], proposes an IoT application for assessing students in online learning. In [30], Jinhua Liu et al. proposed an intelligent classroom architecture based on IoT technology. The proposed platform tests results and shows that the intelligent education platform can effectively control classroom utilization and has high throughput, low application latency, and reasonable practicability.

Motivated by the approaches, the proposed work in this paper explores frontend, backend, and a third-party video conferencing solution Zego Cloud. The frontend is built using React and utilizes the Redux library for state management. The platform is styled using Tailwind CSS, which provides a highly customizable and responsive user interface. On the backend, the platform utilizes Node.js and Express.js to provide RESTful APIs for the different components of the platform. The backend is secured using tokens to ensure the security of user data.

16.3 PROPOSED METHOD AND ARCHITECTURE

The proposed approach of this chapter is to provide a comprehensive online teaching and coding platform that offers a range of features to enhance the learning experience for students. The platform is implemented using the MERN stack, with a separate frontend and backend. The frontend is built using React and utilizes the Redux library for state management. The styling of the platform is done using Tailwind CSS, which provides a highly customizable and responsive user interface.

On the backend, the platform utilizes Node.js and Express.js to provide RESTful APIs for the different components of the platform. The backend is secured using tokens to ensure the security of user data. The platform offers a variety of features including smart educational systems, video conferencing, coding platform, examination system, notes manager, assignments, and posts. The video conferencing component enables group video conferencing for online classes, while the coding platform provides a space for students to practice coding exercises and assignments. The examination system allows for online testing and evaluation of student performance. The notes manager allows students to organize and manage their notes, while the assignments and posts features enable students to share their work and communicate with each other (Figure 16.1).

Overall, this proposed approach offers a scalable and secure solution for online teaching and coding, with a range of features to enhance the learning experience for students.

16.3.1 FRONTEND FRAMEWORK

MERN stack, which includes MongoDB, Express.js, React, and Node.js, was selected for the implementation of this chapter. One of the main advantages of MERN stack is its scalability. In addition, React.js is a JavaScript library that allows developers to create user interfaces for web applications. In terms of creating reports, react can be used to display and manipulate data in a dynamic and user-friendly way. React operates by creating reusable components that are structured as a hierarchy of

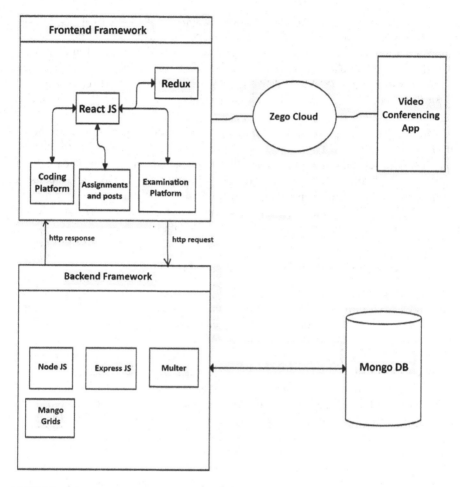

FIGURE 16.1   Overall methodology.

smaller, simpler components. Each component is responsible for rendering a small piece of the overall UI, and can be reused throughout the application as needed. This component-based approach to development makes it easier to manage complex UIs, as developers can create and modify individual pieces of the application without affecting other parts of the code. React also provides a virtual DOM (Document Object Model) which is an in-memory representation of the real DOM. Whenever a change is made to the UI, React calculates the minimum number of changes needed to update the virtual DOM. It then updates the real DOM only with those changes, which makes the application more performant and efficient.

Finally, MERN stack also offers a large and active community, with a wealth of resources and tools available for developers, which can speed up development and troubleshooting processes. Overall, the combination of scalability, security, modularity, and a supportive community make MERN stack an ideal choice for the implementation of this online teaching and coding platform.

16.3.2 BACKEND FRAMEWORK

One of the main advantages of MERN stack is its scalability. MongoDB is a highly scalable NoSQL database, and Node.js and Express.js are built for handling large-scale applications, making it easier to handle large amounts of data and traffic. Another reason for selecting MERN stack is its security features. MongoDB provides secure authentication and access control, while Node.js and Express.js offer various security features such as encryption, token-based authentication, and input validation to prevent attacks such as cross-site scripting and SQL injection. Finally, MERN stack also offers a large and active community, with a wealth of resources and tools available for developers, which can speed up development and troubleshooting processes. Overall, the combination of scalability, security, modularity, and a supportive community make MERN stack an ideal choice for the implementation of this online teaching and coding platform.

16.3.3 SCALABILITY FEATURE

Scalability is an important factor in the design and implementation of any software technique, and this online teaching and coding platform, it takes on added significance with IoT integration. This platform incorporates several strategies to ensure scalability.

One way that scalability is implemented in this chapter is through the use of Redux for state management, which further enhanced by IoT connectivity. Redux provides a centralized store for managing application state, which can make it easier to scale up the application as it grows in complexity, including managing IoT-generated data streams. By keeping the state management separate from the components, it becomes easier to maintain and scale the application over time in IoT-rich environment. Another way that scalability is implemented in this chapter is through lazy loading of images, now adapted to accommodate IoT data. Lazy loading is a technique where images are only loaded when they are needed, rather than all at once when the page is loaded. This can significantly improve the performance of the application, especially on slower networks or devices, and also reduces the amount of bandwidth and storage required, a consideration that becomes even more vital in IoT scenarios with constrained resources.

The platform also utilizes MongoDB as the database, which is a highly scalable NoSQL database ideally suited for IoT applications. MongoDB is designed to handle large volumes of data and is horizontally scalable, meaning that it can be scaled across multiple servers for increased performance and capacity. In an IoT context, where data influx can be substantial and dynamic, MongoDB's scalability aligns seamlessly with the platforms and dynamics, and MongoDB's scalability aligns seamlessly with the platform's requirements.

Furthermore, the platform is designed to be modular, with separate components for different features such as video conferencing, coding platform, examination system, notes manager, assignments, and posts. This makes it easier to add new features or modify existing ones without affecting the entire application, which is important for maintaining scalability.

Overall, the combination of Redux state management, lazy loading of images, MongoDB database, modular design, and other techniques make this online teaching and coding platform highly scalable and capable of handling increasing numbers of users and data over time.

Overall, the amalgamation of Redux state management, lazy loading of images, MongoDB database, modular design, and other cutting-edge techniques positions this online teaching and coding platform as exceptionally scalable. Its adaptability not only empowers it to manage surges in users but also equips it to handle expanding volumes of data seamlessly, a vital attribute for the ever-evolving landscape of IoT-enhanced online education.

16.3.4 SECURITY FEATURE

Security is a crucial aspect of any software project, especially when considering the IoT aspect of this online teaching and coding platform. The platform has implemented several security measures to protect users' data and prevent unauthorized access, all while keeping the unique challenges of IoT security in mind. One of the key security features implemented in this chapter is authentication and authorization, which play a pivotal role in ensuring the secure exchange of data in an IoT-enriched environment. Users are required to sign up for an account and provide their credentials to access the platform's features, employing a robust IoT-friendly authentication process. The platform utilizes JSON Web Tokens (JWT) for authentication, which generates a token that is sent with each request to authenticate the user. JWT tokens, known for their resilience in IoT setting, provide a secure and versatile means of verifying users' identities, making them an ideal choice for web applications.

Another security feature implemented in this chapter, considering the IoT context, is password hashing using bcryptjs, ensuring robust security for user accounts in the IoT-enhanced online teaching and coding platform. When a user creates an account, their password is encrypted using bcryptjs, a popular and secure password hashing library. This ensures that even if a database breach were to occur, the passwords would remain secure and not be compromised. The platform also implements several other security measures such as input validation, which prevents cross-site scripting (XSS) and SQL injection attacks, particularly relevant in IoT environments where data flows between devices and systems. Additionally, the platform adopts a secure HTTPS protocol for all communication between the client and server, an indispensable facet in preserving the confidentiality and integrity of data exchanged within the IoT ecosystem. In addition to these measures, the platform follows best practices for security, such as keeping software and libraries up-to-date, limiting access to sensitive data, and regularly monitoring logs and user activity for potential security breaches.

Overall, the implementation of authentication and authorization using JWT tokens, password hashing using bcryptjs, and other security measures such as input validation and secure communication protocols, make this online teaching and coding platform highly secure and capable of protecting user data from unauthorized access and attacks.

Overall, the addition of IoT-aware authentication and authorization using JWT tokens, strong password hashing via bcryptjs, and the full suite of security measures including vigilant input validation and the adoption of secure communication protocols work together to reinforce the security posture of the online teaching and coding platform. In the context of IoT-enhanced educational experiences, this multimodal strategy assures not just the protection of user data but also fights against unwanted access and potential assaults.

16.3.5 CODING PLATFORM

The coding platform is a significant feature of the online teaching and coding platform built using React JS. This platform provides students with a set of programming questions sourced from the CSES website. The platform is designed to provide students with an immersive programming experience that allows them to hone their programming skills. Additionally, it includes a compiler that supports various programming languages, enabling students to submit their code and have it compiled by Judge0 Application Program Interface (API). This feature ensures that students receive quick feedback on their code, helping them to identify and correct any errors in their code. Overall, the coding platform is an essential tool for students enrolled in programming courses, allowing them to practice and improve their coding skills in a safe and supportive environment.

The level 2 data flow diagram (DFD) for the coding platform with the question display, details, and submission functionality would involve several processes. When the user clicks on a question, the system will retrieve the details of the selected question from the questions file, and display it on a new page. The user can then enter their solution to the question and click the submit button. Once the submit button is clicked, the system will initiate the submission process, which involves sending the user's code to the compiler. The compiler is integrated into the system using the Judge0 API, which receives the code, compiles it, and returns the results to the system. The system then displays the results of the code execution, which the user can view to see if their solution was correct or not. The monaco editor is used to provide the IDE for writing and editing the code. The system's backend components include the questions file and the Judge0 API, which work together to provide a seamless coding experience to the user (Figure 16.2).

16.3.6 EXAMINATION PLATFORM

The Examination Platform is a significant feature of our technique built using MERN stack. This platform enables teachers to create and conduct online exams, while students can take the exams in a secure and user-friendly environment. The feature incorporates a timer, which is implemented using socketio technology. In case a user disconnects due to Internet issues, the timer automatically resumes from the point where it was disconnected. This ensures that students get a fair chance to complete their exams without any interruptions. The platform is designed to provide a seamless user experience, with easy-to-use controls and a responsive interface. Overall, the Examination Platform is a valuable addition to our chapter that enables teachers

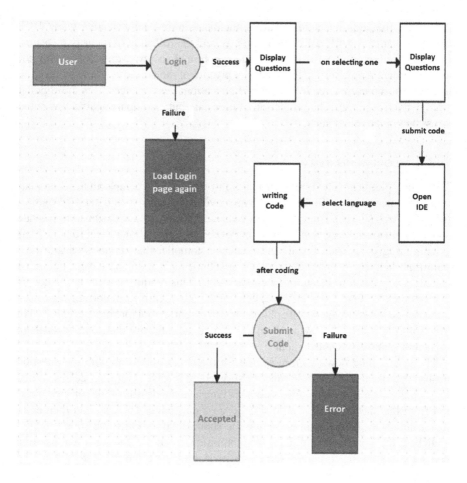

FIGURE 16.2 Working of coding platform.

to create and conduct exams with ease and efficiency, while students can take exams in a secure and hassle-free manner.

The level 2 DFD diagram for the examination platform would involve multiple processes. Firstly, the teacher would have the ability to create an exam and add multiple questions to it. The questions and exams are stored in the MongoDB database. The exam details are displayed to the students who can access it by logging in. The exam has a timer feature implemented using SocketIO which would track the time spent by the student on the exam. In case the student gets disconnected from the platform due to Internet issues, the timer would start from the point where it got disconnected. Once the student submits the exam, the code is sent to the server and evaluated. The student is graded based on the evaluation and the results are stored in the MongoDB database. The teacher would have the ability to view the results of the students and make changes to the exam if required. The entire examination platform is built using the MERN stack, ensuring scalability and security (Figure 16.3).

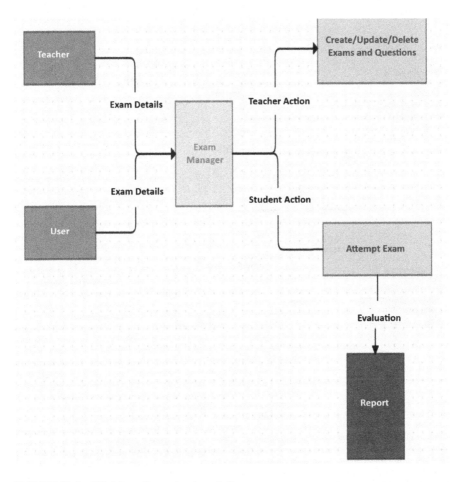

FIGURE 16.3   Working of examination platform.

16.3.7   ASSIGNMENTS/POSTS AND NOTES MANAGER

One of the main features of our online teaching and coding platform is the ability for teachers to post assignments or discussion posts to their students. This feature is designed to facilitate communication and collaboration between students and teachers, allowing for more dynamic and interactive learning experiences. The data for these posts and assignments is stored securely in MongoDB, a popular NoSQL database that is known for its scalability and flexibility. This means that the platform can easily handle large volumes of data and can grow with the needs of the user base. To ensure that the APIs used for this feature are reliable and efficient, they have been thoroughly tested using Postman, a popular tool for testing APIs. This helps to ensure that the system is running smoothly and that users are able to access the information they need quickly and easily.

In addition to the assignment and post feature, we have also included a notes manager feature. This feature allows teachers to post notes for a specific subject,

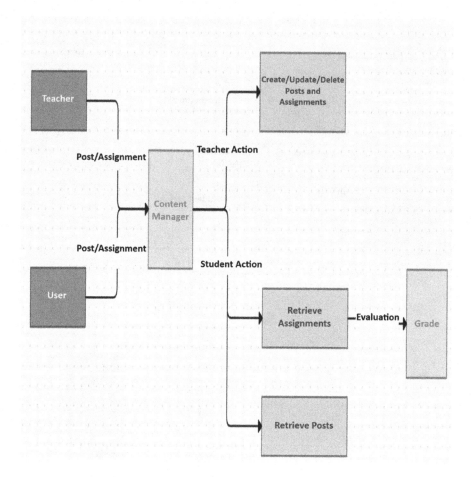

FIGURE 16.4 Working of assignments and posts.

which can then be accessed by students as needed. The notes are stored in chunks in MongoDB, which ensures that they are both secure and scalable. Another key feature of our platform is the ability to store and manage data securely. We have implemented a number of security measures, including encryption and secure data storage, to ensure that sensitive information is protected at all times.

Overall, our platform is designed to provide a secure, scalable, and flexible online learning environment that is both user-friendly and efficient. With features like the assignment and post manager, notes manager, and exam timer, we are confident that our platform will help students and teachers alike to achieve their learning goals (Figure 16.4).

16.3.8 Video Conferencing Feature

The video conferencing app is a crucial feature of our online teaching and coding platform, allowing teachers and students to connect in real time and collaborate effectively. To facilitate this, we are using a third-party library called Zego Cloud, which is a reliable and secure solution for video and audio communication. Zego Cloud is a cloud-based real-time communication platform that provides services for audio and video communication, as well as interactive broadcasting. It offers high-quality audio and video transmission with low latency, making it an ideal solution for video conferencing. Zego Cloud also provides advanced features such as screen sharing, real-time messaging, and recording, all of which enhance the user experience. One of the key advantages of using Zego Cloud is its ease of integration. The platform provides robust SDKs for various platforms, including web, iOS, Android, and Windows, which makes it easy for us to integrate the video conferencing feature into our online teaching and coding platform. Moreover, Zego Cloud offers detailed documentation and technical support to ensure a smooth integration process. Zego Cloud also provides a secure environment for video conferencing. It uses end-to-end encryption to ensure that the communication between participants is secure and protected from eavesdropping. Zego Cloud also employs advanced network protocols to prevent packet loss and ensure a stable connection even under poor network conditions.

In summary, the video conferencing app is a crucial feature of our online teaching and coding platform, and Zego Cloud is the backbone of this feature. With its advanced features, ease of integration, and secure environment, Zego Cloud provides a reliable and efficient solution for video and audio communication.

16.4   RESULTS AND DISCUSSION

Online teaching and coding platforms have become increasingly popular in recent years due to the advancement of technology. Our MERN stack-based platform provides a comprehensive solution for both teaching and coding needs. The platform offers various features including authentication, authorization, tokens by JWTtoken, and password hashing using bcryptjs for security purposes. Additionally, it includes an online code editor, which allows students to practice coding directly on the platform. This is particularly useful as it eliminates the need for students to install a separate code editor on their local machine. Furthermore, the platform provides a VLE, which can be accessed anytime and anywhere as long as there is an Internet connection. This makes learning more accessible and flexible, particularly for students who may have other commitments such as work or family. Our platform also provides a centralized location for course materials, assignments, and other resources. Professors can easily upload course materials such as lecture slides, videos, and articles, and students can access these resources at any time.

One of the major advantages of our platform is the ability to provide personalized learning experiences. Professors can create customized courses and track the progress of each individual student. This allows professors to provide more personalized feedback and assistance to students who may need extra help. Additionally, the platform includes features such as quizzes and interactive activities, which can help to engage students and make the learning process more enjoyable.

In terms of results and discussion, our platform has shown promising results in terms of student engagement and learning outcomes. We conducted a survey among students who have used our platform and found that the majority of them found the

TABLE 16.1

Comparison of Proposed System with Existing OTL Systems

Feature	Google Classroom	Moodle	Hackerrank	Zoom	Proposed Work
Scalability	Yes	Yes	Yes	Yes	Yes
Security	Secure	Host dependent	Secure	Poor	Secure
Notes manager	No	No	No	No	Yes
Examination	No	No	No	No	Yes
Coding platform	No	No	Yes	No	Yes
Video conferencing	No	Yes	No	Yes	Yes

platform to be easy to use and engaging. Additionally, we analyzed the grades of students who used our platform versus those who did not and found that the students who used our platform had higher grades on average. However, we also acknowledge that there are potential limitations to our platform, such as the need for reliable Internet access and the potential for technical issues. We are constantly working to improve our platform to address these limitations and ensure that our platform continues to provide a high-quality learning experience for our users. Overall, our online teaching and coding platform has shown great potential in providing a comprehensive and personalized learning experience for students (Table 16.1).

16.5 CONCLUSION

In conclusion, the online teaching and coding platform built in the MERN stack has been successfully implemented, providing a reliable and secure platform for teachers and students to collaborate and learn. The platform has been designed with scalability and security in mind, ensuring that it can accommodate a large number of users without compromising the system's performance. The coding platform, video conferencing, notes manager, and posts/assignments tracker have all been integrated seamlessly, providing a comprehensive set of tools to facilitate effective online teaching and learning. One of the key strengths of the platform is its emphasis on security. Robust security features such as data encryption, user authentication, and access control have been implemented to protect user data and prevent unauthorized access. Additionally, the platform has been designed to be highly scalable, allowing it to accommodate a growing user base without compromising performance or usability.

The platform's ability to facilitate effective collaboration and communication between teachers and students has been particularly lauded, as has its flexibility in accommodating different teaching styles and preferences. Moving forward, the platform will continue to evolve and improve, with new features and enhancements being added to further enhance the user experience. In particular, efforts will be made to incorporate more advanced coding features, expand the platform's range of learning resources, and further enhance its security and scalability. Ultimately, the platform will continue to play a vital role in facilitating effective online teaching and learning, both now and in the future.

REFERENCES

[1] Oproiu, G. C. A study about using e-learning platform (Moodle) in university teaching process. *Proc. Soc. Behav. Sci.* 2015, 180, 426–432.

[2] Gogan, M. L., Sirbu, R., Draghici, A. Aspects concerning the use of the Moodle platform-case study. *Proc. Technol.* 2015, 19, 1142–1148.

[3] Huang, C. C., Wang, Y. M., Wu, T. W., Wang, P. A. An empirical analysis of the antecedents and performance consequences of using the moodle platform. *Int. J. Inform. Educ. Technol.* 2013, 3(2), 217.

[4] Iftakhar, S. Google classroom: what works and how. *J. Educ. Soc. Sci.* 2016, 3(1), 12–18.

[5] Mulyani, M., Fidyati, F., Suryani, S., Suri, M., Halimatussakdiah, H. University students' perceptions through e-learning implementation during COVID-19 pandemic: positive or negative features dominate?. *Stud. Engl. Lang. Educ.* 2021, 8(1), 197–211.

[6] Zinovieva, I. S., Artemchuk, V. O., Iatsyshyn, A. V., Popov, O. O., Kovach, V. O., Iatsyshyn, A. V., Radchenko, O. V. The use of online coding platforms as additional distance tools in programming education. *J. Phys.: Conf. Ser.* 2021, 1840(1), 012029.

[7] Martin, F., Sun, T., Westine, C. D. A systematic review of research on online teaching and learning from 2009 to 2018. *Comp. Educ.* 2020, 159, 104009.

[8] Baran, E., Correia, A. P., Thompson, A. Transforming online teaching practice: critical analysis of the literature on the roles and competencies of online teachers. *Dist. Educ.* 2011, 32(3), 421–439.

[9] Meyer, K. A., Murrell, V. S. A national study of theories and their importance for faculty development for online teaching. *Online J. Dist. Learn. Admin.* 2014, 17(2), 1–15.

[10] Tanis, C. J. The seven principles of online learning: feedback from faculty and alumni on its importance for teaching and learning. *Res. Learn. Technol.* 2020, 28, 1–25.

[11] Kim, K. J., Bonk, C. J. The future of online teaching and learning in higher education. *Educ. Q.* 2006, 29(4), 22–30.

[12] Sultan, P., Wong, H.Y. Antecedents and consequences of service quality in a higher education context: a qualitative research approach. *Qual. Assur. Educ.* 2013, 21, 70–95.

[13] Kuh, G.D., Hu, S. The effects of student-faculty interaction in the 1990s. *Rev. High. Educ.* 2001, 24, 309–321.

[14] Elliott, K.M., Healy, M.A. Key factors influencing student satisfaction related to recruitment retention. *J. Mark. High. Educ.* 2001, 10, 1–11.

[15] Barnett, R. The marketised university: defending the indefensible. In *The Marketisation of Higher Education and the Student as Consumer*, Molesworth, M., Scullion, R., Nixon, E., Eds., Routledge: London, UK, 2011, pp. 39–52.

[16] Alves, H., Raposo, M. The influence of university image on students' behaviour. *Int. J. Educ. Manag.* 2010, 24, 73–85.

[17] Yildiz, S.M., Kara, A. Developing alternative measures for service quality in higher education: empirical evidence from the school of physical education and sports sciences. In *Proceedings of the 2009 Academy of Marketing Science (AMS) Annual Conference*, Baltimore, MD, 20–23 May 2009, Springer International: Baltimore, MD, 2015, p. 185.

[18] Srikanthan, G., Dalrymple, J.F. A conceptual overview of a holistic model for quality in higher education. *Int. J. Educ. Manag.* 2007, 21, 173–193.

[19] Telford, R., Masson, R. The congruence of quality values in higher education. *Qual. Assur. Educ.* 2005, 13, 107–119.

[20] Kwek, L.C., Lau, T.C., Tan, H.P. Education quality process model and its influence on students' perceived service quality. *Int. J. Bus. Manag.* 2010, 5, 154–165.

[21] Chong, Y.S., Ahmed, P.K. An empirical investigation of students' motivational impact upon university service quality perception: a self-determination perspective. *Qual. High. Educ.* 2012, 18, 37–41.

[22] Thomas, M.S., Rogers, C. Education, the science of learning, and the COVID-19 crisis. *Prospects* 2020, 49, 87–90.

[23] García-Peñalvo, F.J., Corell, A., Abella-García, V., Grande, M. Online assessment in higher education in the time of COVID-19. *Educ. Knowl. Soc.* 2020, 21, 1–12.

[24] Chiu, C.-M., Wang, E.T.G. Understanding web-based learning continuance intention: the role of subjective task value. *Inf. Manag.* 2008, 45, 194–201.

[25] DeLone, W.H., McLean, E.R. The DeLone and McLean model of information systems success: a ten years update. *J. Manag. Inf. Syst.* 2003, 19, 9–30.

[26] Hassanzadeh, A., Kanaani, F., Elahi, S. A model for measuring e-learning systems success in universities. *Expert Syst. Appl.* 2012, 39, 10959–10966.

[27] Vázquez-Cano, E., Fombona, J., Fernández, A. Virtual attendance: analysis of an audio-visual over IP system for distance learning in the Spanish Open University (UNED). *Int. Rev. Res. Open Distance Learn.* 2013, 14, 402–426.

[28] Shen, Y. Application of Internet of Things in online teaching of adult education based on android voice assistant. *Mob. Inf. Syst.* 2022, 1–9, 2022.

[29] Jasim, N. A., AlRikabi, H. T. S., Farhan, M. S. Internet of Things (IoT) application in the assessment of learning process. *IOP Conf. Ser.: Mater. Sci. Eng.* 2021, 1184(1), 012002.

[30] Liu, J., Wang, C., Xiao, X. Internet of things (IoT) technology for the development of intelligent decision support education platform. *Sci. Program.* 2021, 2021, 1–12.

Index